electron microscopy
Volume 7

DATE DUE FOR RETURN

Practical Methods in
ELECTRON MICROSCOPY

Volume 7

Edited by
AUDREY M. GLAUERT
Strangeways Research Laboratory
Cambridge

NORTH-HOLLAND PUBLISHING COMPANY
AMSTERDAM · NEW YORK · OXFORD

IMAGE ANALYSIS, ENHANCEMENT AND INTERPRETATION

D.L. MISELL

National Institute for
Medical Research
Mill Hill, London

NORTH-HOLLAND PUBLISHING COMPANY
AMSTERDAM · NEW YORK · OXFORD

Published by:

ELSEVIER NORTH-HOLLAND BIOMEDICAL PRESS
335 JAN VAN GALENSTRAAT. P.O. BOX 211
AMSTERDAM. THE NETHERLANDS

1st edition 1978
2nd printing 1990

Sole distributors for the U.S.A. and Canada:

ELSEVIER NORTH-HOLLAND INC.
52 VANDERBILT AVENUE
NEW YORK. N.Y. 10017

North-Holland Series ISBN 0 7204 4250 8
This book ISBN 0 7204 0666 8

This book is the laboratory edition of Volume 7 of the series 'Practical Methods in Electron Microscopy'.

Printed in The Netherlands

Editor's preface

Electron microscopy is now a standard technique with wide applications in all branches of Science and Technology, and every year a large number of students and research workers start to use the electron microscope and require to be introduced to the instrument and to the techniques for the preparation of specimens. Many books are available describing the techniques of electron microscopy in general terms, but the authors of Practical Methods in Electron Microscopy consider that there is an urgent need for a comprehensive series of laboratory handbooks in which all the techniques of electron microscopy are described in sufficient detail to enable the isolated worker to carry them out successfully. The series of books will eventually cover the whole range of techniques for electron microscopy, including the instrument itself, methods of specimen preparation in biology and the materials sciences, and the analysis of electron micrographs. Only well-established techniques which have been used successfully outside their laboratory of origin will be included.

Great care has been taken in the selection of the authors since it is well known that it is not possible to describe a technique with sufficient practical detail for it to be followed accurately unless one is familiar with the technique oneself. This fact is only too obvious in certain 'one author' texts in which the information provided quickly ceases to be of any practical value once the author moves outside the field of his own experience.

Each book of the series will start from first principles, assuming no specialist knowledge, and will be complete in itself. Following the successful innovation, made by the same publishers in the parallel series Laboratory

Editor's preface

Techniques in Biochemistry and Molecular Biology (edited by T. S. Work and E. Work), each book will be included, together with one or two others of the series, in a hardback edition suitable for libraries and will also be available in an inexpensive edition for individual use in the laboratory. Each book will be revised, independently of the others, at such times as the authors and editor consider necessary, thus keeping the series of books continuously up-to-date.

Strangeways Research Laboratory
Cambridge, England

AUDREY M. GLAUERT, SC. D.
General editor

Author's preface

Beyond a visual examination of an electron micrograph, there are several methods of further analysing the image. This book aims to provide the necessary practical and mathematical background for the analysis of an electron microscope image in order to extract the maximum amount of structural information. Image analysis provides a quantitative way of assessing image defects (defocus, astigmatism) and image resolution. Also the state of preservation of the specimen can be assessed if it consists of an ordered array of subunits. On the basis of this analysis, structural and non-structural (noise) information may often be separated to give an enhanced image, which shows more clearly the structure of the specimen.

Instrumental methods of image enhancement are also described; these include the use of the energy-selecting electron microscope, and the scanning transmission electron microscope, which are specialised instruments or techniques. Non-standard imaging techniques such as the use of a long focal length objective lens, dark-field microscopy and single-sideband holography (optical shadowing) are easily achieved in the standard electron microscope.

The problems of image interpretation are considered with particular reference to the limitations imposed by radiation damage and specimen thickness. Only in favourable conditions can the normal two-dimensional image be interpreted in terms of a three-dimensional structure. A brief survey is given of the methods for producing a three-dimensional structure from a series of two-dimensional projections. However, in this book the emphasis is on the analysis, processing and interpretation of the two-dimensional projection of a structure.

This book is intended for the use of the biologist or materials scientist who wishes to improve the quality and interpretation of his electron micrographs, but who has little or no experience of image analysis or image processing. Although there is a substantial mathematical content in this book, it should be possible for non-mathematical scientists to understand the principles of image analysis and processing and be able to assess whether these techniques can be useful in their work.

Hamlet: Do you see yonder cloud that's almost in the shape of a camel?

Polonius: By th'mass and 'tis, like a camel indeed.

Hamlet: Methinks it is like a weasel.

Polonius: It is backed like a weasel.

Hamlet: Or like a whale?

Polonius: Very like a whale.

from *Hamlet* by William Shakespeare

Titles of volumes published in this series:

Acknowledgements

I should like to thank Drs. U. Aebi, P.R. Smith and A.C. Steven of the Biozentrum, University of Basel, for their advice, encouragement and preparation of several photographs specifically for this book. Many scientists provided original photographs used in this book, and I am particularly grateful to Professor C.A. Taylor and to Bell and Hyman Ltd. for permission to use optical transforms from '*Atlas of Optical Transforms*' by G. Harburn, C.A. Taylor and T. Welberry.

I am grateful to the following copyright holders for permission to reproduce the illustrations indicated: Academic Press Inc. (*Journal of Molecular Biology*; Figs. 2.1, 3.28, 3.39, 3.41, 4.20, 4.24, 4.25, 4.26, 7.1, 7.2, 7.3, 7.11 and 7.12) (*Journal of Ultrastructure Research*; Figs. 4.9, 4.10, 4.11, 4.12, 4.14, 4.15, 4.16, 5.9, 5.14, 5.18 and 5.19); Alan R. Liss Inc. (*Journal of Supramolecular Structure*; Figs. 3.24, 4.18, 4.19 and 4.20); North-Holland Publishing Company (*Ultramicroscopy*; Fig. 4.7); The Rockefeller University Press (*Journal of Cell Biology*; Figs. 4.12 and 4.13); Nature (*Nature*; Figs. 4.22, 7.9 and 7.10); The Royal Society (*Philosophical Transactions of the Royal Society*; Figs. 4.27, 4.28 and 5.15); The American Association for the Advancement of Science (*Science*; Fig. 4.30); The Royal Microscopical Society (*Journal of Microscopy*; Figs. 4.34, 5.1, 5.2, 5.3, 5.4, 5.5, 5.7 and 5.16); Pergamon Press (*Micron*; Figs. 5.23, 5.24 and 5.25); The Institute of Physics (*Journal of Physics*; Figs. 6.3 and 7.5); and the Company of Biologists (*Journal of Cell Science*; Fig. 7.13).

I am indebted to Elaine Brown for the many hours she spent preparing the diagrams and prints for this book, and to Elaine and my wife, Janet, for preparing the typescript.

London, April 1978   D.L. MISELL

Contents

Chapter 6. Computer methods of contrast enhancement 245

Chapter 7. Image interpretation 259

Chapter 1

Introduction

Electron microscopy is a visual science, but observation of the image or the electron micrograph does not necessarily lead to the extraction of the maximum amount of available information. By using the techniques detailed in this book, it should be possible to extract all the structural information that is present in the electron micrograph and to quantify it.

Biological materials are prepared in several different ways in the hope of finding a method which most nearly preserves the original structure. For example, in the examination of a system such as the head of the bacteriophage T4 (Aebi et al. 1976, 1977), one of the objectives is to determine the geometrical arrangement and shape of the protein molecules in the capsid. Negatively-stained, freeze-dried, freeze-fractured or metal-shadowed preparations may be used, together with various imaging techniques that minimise radiation damage to the specimen. It is virtually impossible, however, to assess quantitatively the differences between the images obtained without using additional techniques such as optical diffractometry (Horne and Markham 1972) and image reconstruction (Aebi et al. 1976, 1977). For T4 phage heads the fact that the capsid is an assembly of regularly arranged protein subunits enables a determination of their geometrical arrangement to be made, whilst the highest order diffraction spot observed in the *optical transform* (§ 3.3), or *diffractogram*, gives some indication of the order preserved in the specimen. In addition, in such preparations, the images of the top and bottom surfaces are superimposed. This leads to an uninformative moiré pattern, but careful analysis of the optical transform of the image allows the contributions of the two surfaces to be separated.

The next step is to use the optical transform to produce an *enhanced image*, based on the regular (periodic) arrangement of protein subunits. More accurately this procedure should be referred to as *image reconstruction,* image enhancement being the rebuilding of the original micrograph from only that information which is relevant to the specimen structure. Image enhancement can also be produced using computer based analysis of a numerical *Fourier transform* (§ 3.4). In this procedure the optical density variations in the micrograph are converted to numbers for *digital processing.* This has some distinct advantages; for example, the image can be corrected for small distortions in the positions of subunits nominally in a regular array (Crowther and Sleytr 1977). Both optical and digital processing of electron images provide quantitative information on how well the biological structure has been preserved, the artefacts introduced by the techniques used to prepare the specimen, and, in high resolution applications (0.5–2.0 nm), the effects of radiation damage.

Image reconstruction may be practical in only a few laboratories, but the analysis of images by optical diffractometry should be a routine procedure for all microscopists who wish to make the best use of their micrographs, irrespective of specimen resolution. The basic principles of image analysis by both optical and computer methods are discussed in Chapter 3, and some typical applications in materials science and biology are described.

It will be evident that image analysis is most useful when the specimen has some type of symmetry, such as translational (as in a two-dimensional array of units), helical or rotational. This is not at all uncommon in biology. A number of membrane systems show a regular arrangement of protein subunits, for example gap junctions and bacterial outer membranes (Thornley et al. 1974). Also, some viruses, like adenovirus and T-even phages, show a periodic arrangement of subunits. Helical symmetry is found in microtubules, tobacco mosaic virus (TMV), T4 phage tails and in some nucleoprotein complexes from viruses, such as Sendai virus. Rotational symmetry, where a repetition of the structure occurs on rotating the particle about its centre through $360°/n$ for n-fold symmetry, is exhibited by T4 phage base plates ($n = 6$), TMV stacked disc protein ($n = 17$), and assemblies of proteins from virus surfaces such as the 'groups of nine' hexons of adenovirus ($n = 3$).

Frequently the order within the specimen is only preserved over a small number of subunits, but this is sufficient for at least a preliminary structural analysis by electron microscopy. Certainly, if large (≥ 200 μm) three-dimensional crystalline arrays are available, electron microscopy cannot compete with X-ray crystallography, but often such crystals cannot be made. For example, the purple membrane of *Halobacterium halobium* occurs in two-dimensional, highly ordered sheets; only limited information is available using X-ray diffraction techniques (Blaurock 1975; Henderson 1975), but careful (low radiation dose) electron microscopy of unstained purple membrane sheets and image reconstruction techniques have led to a three-dimensional structure of the membrane protein at 0.7 nm resolution (Henderson and Unwin 1975; Unwin and Henderson 1975).

In addition to naturally occuring ordered systems, there are several examples where, by biochemical methods, normally irregular (aperiodic) structures can be encouraged to form quite well ordered crystalline arrays; examples include neuraminidase and haemagglutinin, proteins from the envelope of influenza virus. Whole viruses may also be encouraged to form large two-dimensional ordered arrays (Horne and Ronchetti 1974).

Even if the specimen shows no order, image analysis provides a rapid way of assessing image quality; that is, how good it is electron-optically, with respect to focus, astigmatism, specimen drift and other image defects (§ 3.3). Of course, the microscopist tries to minimise these defects by correct use of the microscope, but image analysis (particularly optical diffractometry) will provide a rapid post-diagnostic measurement of the magnitude of image defocus and other defects.

Most image reconstruction techniques are designed for specimens exhibiting symmetry (see Chapter 4), but there are several methods of image enhancement for aperiodic or amorphous specimens. These are more subjective than the methods for ordered specimens, but they have sometimes led to a gain in structural information. For example, in attempting to resolve single heavy atoms in labelled biological macromolecules, such as deoxyribonucleic acid (DNA), and organo-metallic compounds, *spatial (Fourier) filtering* (§ 4.7) is used to reduce the high frequency noise in the image and facilitate the determination of atom positions (Ottensmeyer et al. 1972). At a much lower resolution the enhancement of boundary structures in sectioned material can be achieved by, for example, differentiating the image (§ 6.3).

Image processing is the next step to follow image analysis. Its objective is to produce a structurally enhanced version of the original image. It is not a cosmetic for making poor images into good ones, but it does enable the extraction of the maximum amount of structural information from the original image. It is no substitute for the use of the best specimen preparation photographed under minimum irradiation conditions, but is a complementary technique. There are published examples where the reconstructed image gives no more information about the specimen structure than was available by a careful examination of the original image. So methods for obtaining the 'best' images (as assessed by optical diffractometry) are emphasized, before proceeding with the long operation of image processing.

The major part of this book is concerned with the analysis and processing of images after they have been photographed by conventional transmission electron microscopy (CTEM), but instrumental methods of improving image contrast are also discussed (Chapter 5). Clearly 'non-conventional' electron microscopy which can produce clearer images is to be preferred to post-processing. Some of these methods require the use of specialist instruments, such as the scanning transmission electron microscope (Crewe and Wall 1970; Crewe 1971) and the energy-selecting electron microscope (Henkelman and Ottensmeyer 1974), while other techniques, such as dark-field electron microscopy, where only a small modification to the conventional electron microscope is required (Kleinschmidt 1971), may be unfamiliar to the majority of biologists.

Assuming that image analysis and image processing have been fully exploited, the final step is to interpret the image in terms of structure, with a consequent inference concerning the biological function of a system which has been examined under the harsh conditions existing in the electron microscope (Chapter 7). However detailed a picture emerges as a result of image processing, the material examined in the electron microscope is in the dried state. Only where other physical techniques confirm the dimensions or arrangement of known structural features can there be said to be a fair understanding of the organisation of the original (and probably hydrated) structure.

The scope of this book excludes a detailed account of one major aspect of image processing, namely, three-dimensional reconstruction techniques (DeRosier and Klug 1968; DeRosier 1971). Only the analysis of two-dimensional projections of a three-dimensional object is included; here the superposition arising from observing a three-dimensional structure in an electron microscope limits the information available (§ 7.3).

The purpose of this book is to convince the biologist that image analysis and image processing or reconstruction offer an additional dimension to electron microscopy, that may lead to structural information over and above that seen directly in the image. So far these techniques have failed to become well established in biological electron microscopy, probably because of the complexity of the mathematics associated with them. Thus in the main part of this book I hope to convince the biologist that image analysis is a valuable tool by showing many practical examples, and providing the mathematical basis as simply as possible.

References

Aebi, U., R.K.L. Bijlenga, B. ten Heggeler, J. Kistler, A.C. Steven and P.R. Smith (1976), A comparison of the structural and chemical composition of giant T-even phage heads, J. Supramol. Struct. *5*, 475–495.

Aebi, U., R. van Driel, R.K.L. Bijlenga, B. ten Heggeler, R. van den Broek, A.C. Steven and P.R. Smith (1977), Capsid fine structure of T-even bacteriophage: binding and location of two dispensable capsid proteins into the P23* surface lattice, J. Mol. Biol. *110*, 687–698.

Blaurock, A.E. (1975), Bacteriorhodopsin: a trans-membrane pump containing α-helix, J. Mol. Biol. *93*, 139–158.

Crewe, A.V. (1971), High resolution scanning microscopy of biological specimens, Phil. Trans. Roy. Soc. Lond. B *261*, 61–70.

Crewe, A.V. and J. Wall (1970), A scanning microscope with 5 Å resolution, J. Mol. Biol. *48*, 375–393.

Crowther, R.A. and U.B. Sleytr (1977), An analysis of the fine structure of the surface layers from two strains of *Clostridia*, including correction for distorted images, J. Ultrastruct. Res. *58*, 41–49.

DeRosier, D.J. (1971), The reconstruction of three-dimensional images from electron micrographs, Contemp. Phys. *12*, 437–452.

DeRosier, D.J. and A. Klug (1968), Reconstruction of three-dimensional structures from electron micrographs, Nature *217*, 130–134.

Henderson, R. (1975), The structure of the purple membrane from *Halobacterium halobium*: analysis of the X-ray diffraction pattern, J. Mol. Biol. *93*, 123–138.

Henderson, R. and P.N.T. Unwin (1975), Three-dimensional model of purple membrane obtained by electron microscopy, Nature *257*, 28–32.

Henkelman, R.M. and F.P. Ottensmeyer (1974), An energy filter for biological electron microscopy, J. Microsc. (Oxford) *102*, 79–94.

Horne, R.W. and R. Markham (1972), Applications of optical diffraction and image reconstruction techniques to electron micrographs, in: Practical methods in electron microscopy, Vol. 1, part 2, A.M. Glauert, ed. (North-Holland, Amsterdam), pp. 327–434.

Horne, R.W. and I.P. Ronchetti (1974), A negative staining-carbon film technique for studying viruses in the electron microscope. I. Preparative procedures for examining icosahedral and filamentous viruses, J. Ultrastruct. Res. *47*, 361–383.

Kleinschmidt, A.K. (1971), Electron microscopic studies of macromolecules without appositional contrast, Phil. Trans. Roy. Soc. Lond. B *261*, 143–149.

Ottensmeyer, F.P., E.E. Schmidt, T. Jack and J. Powell (1972), Molecular architecture: the optical treatment of dark-field electron micrographs of atoms, J. Ultrastruct. Res. *40*, 546–555.

Thornley, M.J., A.M. Glauert and U.B. Sleytr (1974), Structure and assembly of bacterial surface layers composed of regular arrays of subunits, Phil. Trans. Roy. Soc. Lond. B *268*, 147–153.

Unwin, P.N.T. and R. Henderson (1975), Molecular structure determination by electron microscopy of unstained crystalline specimens, J. Mol. Biol. *94*, 425–440.

Chapter 2

Factors to be considered in image interpretation

Before embarking on the procedures of image analysis and processing which are the main subject of this book, it is essential to bear in mind the fact that changes in the specimen structure inevitably occur before the image is recorded. In particular, staining techniques used to improve image contrast in biological specimens limit the accuracy with which the morphology of the specimen can be determined (Unwin 1975), and radiation damage by the electron beam is a fundamental limitation in the observation of specimens, particularly at high resolution (Stenn and Bahr 1970; Beer et al. 1975; Glaeser 1975).

Biological systems are three-dimensional structures and in the electron microscope these are observed as *two-dimensional projections*, frequently distorted by the drying processes involved in their preparation. For a thin specimen (20 nm thick) this projection may be informative, but it will be of limited use, for example, in the examination of a whole virus 50 nm thick. Clearly this problem may be overcome by examining several different projections of the structure by tilting the specimen in a goniometer stage, but a three-dimensional reconstruction from such projections is a far from routine operation, even when the particle shows a high degree of symmetry (Crowther et al. 1970; Crowther 1971; Crowther and Klug 1975). Furthermore, as the specimen thickness increases, an increasing number of electrons are *inelastically scattered* in the specimen and, because these electrons lose energy in the specimen, they cannot be focused by the objective lens. This results in a loss of image resolution as a result of the *chromatic aberration* of the objective lens (Cosslett 1956), in addition to a loss in image contrast.

Most microscopists take considerable care in focusing the image and usually choose 'in-focus' or underfocus conditions for recording images.

Sometimes the choice of objective lens defocus is influenced by the requirement of a visually attractive image. This is a dangerous practice; the fact that
the choice of defocus influences the appearance of an image should serve as
a warning that no image that depends on a microscope adjustment can be
believed. The contrast reversals observed in a specimen as focus is altered
can often lead to an image whose interpretation depends on the particular
defocus chosen (Johnson and Crawford 1973). The choice of image defocus
depends on the resolution expected from a specimen and becomes more
critical as the resolution improves. In principle, all images can be corrected
for the effect of defocus to give a reconstructed image that is independent of
this microscope parameter (see § 4.4), but in many applications a careful
choice of defocus will give an image that is essentially free from imaging
artefacts (see § 3.2).

2.1   Specimen staining

Positive stains bind to components of the specimen so that images of the
stain closely relate to the specimen structure. Negative stains surround the
biological molecule and penetrate into the hydrophilic regions within it
(Horne 1973). It is therefore assumed by implication that the light patches in
the image (dark on the photographic image) surrounded by the dark stain
represent the biological structure. It must be remembered that in negatively-
stained preparations it is the stain distribution that is observed and it is only
by implication that areas of excluded stain represent the biological structure
(Horne 1973). There are doubts about the accuracy with which the negative
stain maps the morphology of the biological structure and there is also
experimental evidence, using image analysis, that the stain migrates and con-
tracts when subject to *non-minimal radiation* conditions (Unwin 1974).
Whereas few will claim a resolution of better than 2 nm for negatively-stained
specimens (Beer et al. 1975), there is some doubt whether it is even an accurate
representation of a biological structure at this resolution because of simultaneous positive and negative staining (Unwin 1975). Whilst raising these
doubts about negative staining, it must be accepted that this often is the only
way of imaging biological structures with a sufficiently high image contrast.
The image analysis procedures described in this book provide valuable
information concerning the effects of the negative-staining procedure on
specimen structure, and this is discussed in the final chapter (§ 7.1).

   If it is possible to obtain large (100×100 subunits) ordered arrays of the
biological material, subminimal radiation images of *unstained*, sugar

supported, preparations can be obtained, which may be analysed (Unwin and Henderson 1975). A sugar solution (glucose or sucrose) can be used in place of stain as a support for the structure – it is believed to take on a glass-like state, replacing water in the specimen in a similar way to a negative stain. Small arrays of unstained specimens (10 × 10) are insufficient for effective image enhancement. Thus in most cases, and certainly for non-crystalline specimens, staining techniques must be used. For this reason it is suggested that the biological structures are also examined using other specimen preparation techniques, such as freeze-fracturing and -etching, or freeze-drying, followed by heavy metal shadowing, in order to provide some check on the structure determined from negatively-stained specimens. To encourage attempts to examine unstained preparations, an *electron diffraction pattern* (Beeston 1972) obtained from unstained purple membranes, with diffraction spots extending to 0.3 nm resolution (Unwin and Henderson 1975), is shown in Fig. 2.1. Electron diffraction is an excellent way of examining specimens with a periodic structure because it provides a rapid visual assessment of how well ordered the specimen is and of the level of radiation damage, before an image is recorded. In unstained ordered specimens the diffraction pattern fades very rapidly at normal illumination levels; in negatively-stained specimens the diffraction pattern persists much longer before fading, although the diffraction orders rarely extend beyond a resolution of 2 nm (about 1/5 of the extent of the diffraction pattern shown in Fig. 2.1).

On most modern electron microscopes it is a relatively simple procedure to switch directly from the image to the diffraction pattern (Beeston 1972), without removing the objective aperture. In this way it has been possible to examine the success of various preparation techniques on the structural order of crystals of neuraminidase, a protein from influenza virus. The presence of a good, sharp electron diffraction pattern indicated success with only one negative stain, little success with glutaraldehyde cross-linking, and none at all, so far, with unstained crystals. This was a relatively rapid assessment, without going to the trouble of recording images. Once a good preparation method has been found by electron diffraction, images can be taken with some confidence that they will be suitable for analysis and processing. Discussion of the reasons for preferring electron images to electron diffraction patterns for structure determination will be found in Chapter 3, where the essential differences between electron and optical diffraction are explained; for the present it should be mentioned that quite large arrays of subunits are required in order to obtain good selected area diffraction patterns (Beeston 1972).

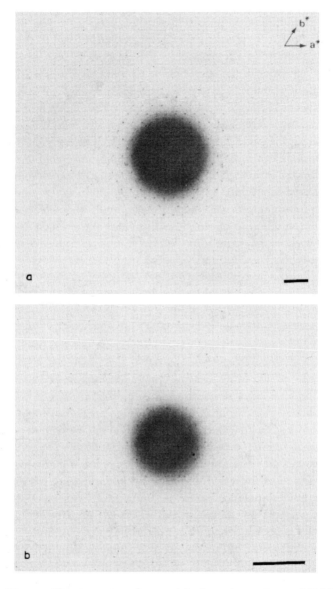

Fig. 2.1. Electron diffraction pattern from unstained purple membrane of *Halobacterium halobium*: (a) high-angle, (b) low-angle patterns. Diffraction bar $= 0.5\,\mathrm{nm}^{-1}$. The highest resolution diffraction spot corresponds to $3\,\mathrm{nm}^{-1}$ or a resolution of $\frac{1}{3} \simeq 0.3\,\mathrm{nm}$. $E_0 = 100\,\mathrm{keV}$. a^* and b^* correspond to the directions of the reciprocal lattice vectors for the hexagonal lattice. (From Unwin and Henderson 1975.)

2.2 Radiation damage

In many non-biological specimens radiation damage is not an important factor, but even here it is recommended that at least some measure of the electron radiation dose is determined. *Radiation dose* at the specimen is measured in terms of the number of electrons n_0 incident on an area of one square nanometre of specimen (n_0 has units $e^- \, nm^{-2}$), where $6.25 \, e^- \, nm^{-2}$ is equivalent to a dose of 1 coulomb per square metre ($C \, m^{-2}$). *Minimal radiation* conditions correspond to the electron dose required to record an image on a photographic emulsion with an optical density of about 1 unit (Williams and Fisher 1970); no significant pre-irradiation of the specimen is allowed (Johansen 1976). For an image recorded at an electron–optical magnification of $50,000 \times$ this corresponds to a radiation dose of $n_0 = 1500$ $e^- \, nm^{-2}$ ($250 \, C \, m^{-2}$) at the specimen for Kodak Electron Image Plates (see Appendix); the dose will vary for different photographic materials and developing conditions. This radiation dose may not seem to be very large, but it should be noted that for electrons of incident energy 100 keV, $1 \, C \, m^{-2}$ $\simeq 400 \, kJ \, kg^{-1} = 40 \, Mrad$, assuming that inelastic electron scattering is totally responsible for radiation damage and that the average energy loss per inelastic collision is 25 eV (see § 2.5). This magnitude of radiation dose is sufficient to denature most proteins in biological specimens. In fact a dose of $1500 \, e^- \, nm^{-2}$ generally exceeds that required for high resolution (0.5 nm) microscopy by a factor of about 20 (Beer et al. 1975; Glaeser 1975; Unwin and Henderson 1975). A *subminimal radiation* dose is defined to be the dose which is insufficient to cause appreciable radiation damage to the specimen. Subminimal radiation corresponds to $n_0 \simeq 50\text{--}100 \, e^- \, nm^{-2}$ ($8\text{--}16 \, C \, m^{-2}$) and will lead to a statistically noisy image of very low optical density with normal photographic emulsions, because there are relatively few electrons incident on the photographic emulsion (see below). The normal dose given to the specimen after irradiation at high magnification ($100,000 \times$), during astigmatism correction and focusing, is about $62,000 \, e^- \, nm^{-2}$ ($10,000 \, Cm^{-2}$)!

Clearly, if there is any indication that the image of a particular specimen depends on the amount of irradiation by the electron beam, an assessment must be made of the radiation dose by determining the *optical density* of the developed photographic emulsion. The number of electrons incident on the specimen will determine the optical density of the photographic image, but the number incident on unit area ($1 \, \mu m^2$) of the emulsion will depend on the *electron-optical magnification, M,* and the resulting optical density will depend on the type of emulsion. The same number of electrons through unit area of the specimen will be spread over four times as large an area of emulsion at a magnification of 100,000 × as compared to 50,000 × ; in order to obtain a developed emulsion with the same optical density, four times as large an electron dose will be required at 100,000 ×. Thus, one way to reduce radiation damage is to take images at as low a magnification as possible. There is a lower limit to M, depending on the resolution of the emulsion as compared to the resolution expected from the specimen, and the accuracy with which focus can be determined on the viewing screen. However, with a specimen exhibiting a resolution of 2 nm, M can be as low as 10,000 × before the resolution of the emulsion (usually about 10 μm) is a limiting factor.

Quantitatively the approximate relationship between optical density D, specimen dose n_0 (e$^-$ nm^{-2}) and magnification M (in units of 1000) is:

$$n_0 = M^2 D/D_0 \qquad (2.1)$$

where D_0 is the optical density on the photographic emulsion for an electron dose 1 e$^-$ μm^{-2} at the image plane. D_0 is normally available from the manufacturer's data for a particular emulsion and *standard* developing conditions. For example, $D_0 = 2$ for Kodak Electron Image Plates using high speed developing conditions (D19 undiluted or HRP 1 : 2 for 5 min) and an incident electron energy of 100 keV. Equation (2.1) can be used in two ways as illustrated in Table 2.1. Firstly, it can be used to determine the maximum optical density D once the specimen dose n_0 and image magnification M are specified for (a) subminimal radiation, or (b) minimal radiation. Evidently using a high magnification for low dose images leads to an image with a very low optical density, which may be visually unacceptable. Such an image may, however, be suitable for image analysis provided that the specimen comprises a large number of regularly arranged subunits (Unwin and Henderson 1975). Secondly, if it is decided that the final photographic image should have an optical density of 1, the magnification chosen determines the radiation dose to the specimen (Table 2.1c). Of course, these calculations assume little or no pre-irradiation of the specimen; focusing and astigmatism correction are carried out on an area of carbon support film adjacent to the area of interest and the specimen is irradiated only for the time required to obtain a photographic record. It is recommended that all images are recorded under measured radiation conditions, so that in any comparison of images obtained from different preparations or on different occasions, this factor is at least known, if not controlled. Provided standard

TABLE 2.1

The relationship between electron dose (n_0 electrons nm^{-2}), image magnification (M in units of 1000) and optical density (D) for Kodak Electron Image Plates developed for maximum emulsion speed

	n_0 (e^-nm^{-2})*	M	D
(a)	100 (subminimal radiation)	10	2.00
	100	20	0.50
	100	30	0.22
	100	40	0.13
	100	50	0.08
	100	100	0.02
(b)	2000 (minimal radiation)	30	4.44
	2000	40	2.50
	2000	50	1.60
	2000	100	0.40
	2000	200	0.10
(c)	50	10	1.00
	200	20	1.00
	450	30	1.00
	800	40	1.00
	1250	50	1.00
	5000	100	1.00
	20,000	200	1.00

* 1 Coulomb $m^{-2} \equiv 6.25$ electrons nm^{-2}
(or 10^{-4} coulomb $cm^{-2} \equiv 6.25 \times 10^{-2}$ electrons Å^{-2})
Incident electron energy $E_0 = 100\,keV$.

developing conditions are used, an estimate of n_0 can always be obtained. For example, if the average optical density of a photograph is measured on a microdensitometer (e.g. Joyce-Loebl) to be 1.5 units at $M = 50,000 \times$, then, using Eq. (2.1) the dose given to the specimen will be $n_0 = 1875\,e^-\,nm^{-2}$. The use of non-standard developing conditions means that it will be impossible to obtain reliable measurements of n_0. More reliable radiation dose measurements can be obtained by using a retractable Faraday cage fitted into the side of the microscope just above the viewing screen (Johansen 1976), but the use of optical density measurements will serve as a routine method.

An illustration of the structural effects of radiation damage is given in Fig. 2.2, which shows a series of images of negatively-stained T-layer from *Bacillus brevis*. These results pre-empt the results of image processing (see

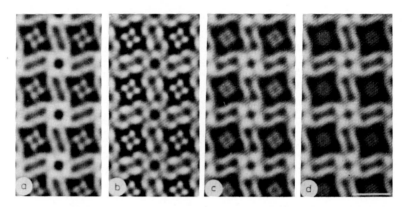

Fig. 2.2.   Bright-field images of negatively-stained (sodium phosphotungstate) T-layer from *Bacillus brevis* after image processing, showing the effect of electron dose on structural information. The protein is shown as white on a dark background of negative stain. Electron dose is (a) $5000 \, e^- \, nm^{-2}$ (minimal radiation), (b) $60 \, e^- \, nm^{-2}$ (subminimal radiation), (c) 62,000 $e^- \, nm^{-2}$ and (d) $620,000 \, e^- \, nm^{-2}$ at the specimen. Image bar = 10 nm, $E_0 = 80 \, keV$. (P.R. Smith and J. Dubochet, unpublished.)

Chapter 4), because the original images have been subjected to filtering to remove non-periodic information (noise) and also to separate the images of the two superimposed surfaces of the T-layer (Aebi et al. 1973). Radiation damage cannot be visually assessed from the original images. In the processed images the protein (regions of excluded stain) is shown as white. It is evident that minimal radiation conditions (Fig. 2.2a, $n_0 = 5,000 \, e^-$ nm^{-2}) give quite a good representation of the arrangement of protein subunits, only slightly inferior to subminimal radiation conditions (Fig. 2.2b, $n_0 = 60 \, e^- \, nm^{-2}$). Large radiation doses (Fig. 2.2c and d), however, lead to a progressive loss of structural information. This assessment of the effect of radiation damage could have been achieved equally well by examining the optical transforms of the original images, in which the diffraction spots would progressively decrease in intensity, with the higher-order (high resolution) spots eventually disappearing (see Chapter 3). From such studies on negatively-stained preparations, it is evident that n_0 should be as low as 1000–2000 $e^- \, nm^{-2}$. Negatively-stained specimens appear to be less radiation sensitive than unstained biological specimens ($n_0 \simeq 50 \, e^- \, nm^{-2}$), but even then the resolution achieved is rarely better than 2 nm. The effects of radiation damage are considered in greater detail in Chapter 7.

Unfortunately photographing images under minimal radiation conditions is not a simple procedure. Whatever method is used a large number of photographs will have to be rejected because of unacceptable specimen drift, astigmatism or non-uniform illumination, and often a poor area of specimen will be recorded, because it cannot be selected in advance. The general principle of minimal irradiation is to avoid irradiating the specimen before you are ready to take the picture. Firstly the microscope must be well aligned at high magnification. The objective aperture may be removed but a small aperture (20–100 μm diameter) should be used in the second condenser lens (C2), and the first condenser lens (C1) should be strongly excited to produce a small patch of illumination. Astigmatism is corrected at a magnification of about 300,000 × using the granular appearance of the phase contrast image of the carbon support film (Agar et al. 1974). This correction may be made before inserting the small C2 aperture because of the low illumination level on the viewing screen. The image magnification is then reduced to the required value (usually 20,000–50,000 ×), and with an expanded illumination spot of very low intensity the specimen grid is moved systematically until a suitable area of specimen enters the field of view. The beam is then switched off, using either an electrical beam deflector system above the specimen or a mechanical shutter at the level of the first condenser lens (C1). A foot-switch is a most convenient control for electrical deflectors, and a simple pair of metal rings around the C1 aperture control, one stopping against the other, ensures that the aperture can be moved to block the beam and then returned to its original position with accuracy. The obstructed beam is now deflected using the electrical dark-field controls which are fitted on most modern microscopes, so that when it is switched on again, the specimen will not be irradiated, but only an adjacent area. The intensity of the illumination can then be increased to a level that is satisfactory for focusing. Once the objective lens has been set for the required amount of defocus, the beam is switched off again, followed by the dark-field tilt. Then the camera shutter is opened so that the photographic film or plate will be exposed as soon as the beam is reinstated. The exposure time is controlled manually using the foot-switch or C1 shutter to turn off the beam again. After completion of the exposure, the camera shutter is closed and examination of the specimen is continued under low illumination conditions. Care must be taken to scan the specimen systematically so as to avoid looking at the same area twice.

These techniques need a great deal of practice in order to set up an efficient routine of picture taking. The dark-field controls must be adjusted so that the beam is not deflected too little (thus irradiating the specimen) or too much (thus losing the beam). There are illumination alignment problems in using a small condenser (C2) aperture; it is possible to use a normal size aperture (200–300 μm) but the size of the illumination spot may be so large that the specimen is pre-irradiated before it is seen on the viewing screen.

There are several variations on this procedure for taking minimal irradiation pictures. For example an off-axis auxiliary viewing screen, about 50 mm in diameter may be fitted at the front of the viewing chamber (as suggested by Unwin 1974). The dark-field tilt electrical deflectors can then be adjusted so that the illumination is focused on this front screen, well away from the 'straight through' beam path. When a good area of specimen is seen at the edge of the front screen, it is assumed that there is more at the centre of the main viewing screen. After focusing on the auxiliary screen, the camera shutter is opened and the dark-field tilt is switched off to start the exposure, and switched on again to end the exposure.

In older electron microscopes many of these electrical deflection systems will be absent, so mechanical methods must be used. The C1 shutter already mentioned can be used, and the specimen moved so that a new area is in the beam path when the shutter is opened. This procedure often gives serious specimen drift (Williams and Fisher 1970). Before rejecting the need for minimal radiation techniques, it is worthwhile trying some of these methods to see if better images are obtained than by using normal procedures (see Williams and Fisher 1970); a quantitative assessment of any improvements in the images of crystalline specimens can be obtained by examining the optical transform (see Chapter 3).

2.3   Specimen thickness

It is important to remember that in electron microscopy a two-dimensional projection of a three-dimensional structure is being viewed. Often this limits the interpretation of an image, particularly if the specimen is thick (e.g. a whole virus), because of the superposition of structural information. Thus, even with the aid of image analysis and image reconstruction techniques, a single view can be only tentatively interpreted in terms of the actual structure of the specimen. The exception is an image of a helical particle which contains different views of the subunits simultaneously because they are arranged on the helix. In all other structures different projections must be obtained by tilting the specimen. Assuming that a particle of diameter D has no symmetry, the number of views N required for a three-dimensional reconstruction at a resolution of d is approximately (Klug 1971):

$$N = \pi D/d \qquad (2.2)$$

Note that for a helical particle D does not refer to the diameter of the whole particle but to the diameter or separation of the subunits forming the helix (see § 3.6).

Thus if the three-dimensional structure of an asymmetric particle 20 nm in diameter (such as a ribosome) is to be determined to a resolution of 2 nm, N is about 30, the views being selected at tilt angles equidistantly spaced in the range $+90°$ to $-90°$. Radiation damage and contamination, and problems like drift of specimen and focus will limit the success of this type of experiment. In general, three-dimensional reconstruction has been limited to highly symmetrical particles, such as spherical viruses (Crowther 1971; Crowther and Klug 1975), for which the number of independent projections required is reduced to a few.

Fortunately, provided that the specimen is thin (20 nm), useful information can still be obtained, for example, on the arrangement of protein subunits in a membrane and on changes in this geometrical arrangement when the biological system is subject to genetic or biochemical variations. Even with thicker specimens, such as crystalline arrays of whole viruses, it may be possible to reduce the effect of superposition by separating the images arising from the top and bottom of the specimen (Klug and DeRosier 1966; see § 3.7).

Two other effects of increasing specimen thickness should also be considered: firstly, as the specimen thickness increases the focus difference between the top and bottom of the specimen increases. Thus, assuming that focus is determined with respect to the carbon support film, the upper and lower surfaces of the specimen will differ in focus by the thickness of the specimen. This effect is important only in high resolution electron microscopy (0.5–2 nm), but it does place a fundamental limit on the interpretation of an image. The second limitation imposed by increasing specimen thickness is based on the assumption that the image is linearly related to the specimen; that is, density variations in the specimen are exactly reproduced in the image. The basic assumptions of image analysis and image reconstruction depend on this linear relation; it is accepted that lens aberrations and other image defects, and the choice of defocus may distort the image but these microscope-dependent factors can be taken into account.

It is not possible to give an exact upper limit to specimen thickness, but it should always be remembered that image reconstruction is based on the image and only by inference does this reconstructed image relate to the specimen structure. This is where other information derived from biochemical or other physical techniques becomes essential; the structure as determined by electron microscopy must be consistent with such additional information.

2.4   Objective lens defocus and lens aberrations

Images are usually photographed just underfocus, often to give the most visually acceptable image. The reason that this is successful is because an underfocus Δf partially cancels the effect of the spherical aberration of the objective lens and leads to minimum distortion of the image and maximum image contrast. However, this is not an objective way of choosing defocus

Fig. 2.3.   Bright-field images of a 50 nm thick transverse section of porcupine quill keratin stained with osmium tetroxide: (a) 'in-focus', (b) defocus $\Delta f = 600$ nm ('9 + 2' model), (c) $\Delta f = 1000$ nm, (d) $\Delta f = 1600$ nm ('5 + 0' model). Image bar = 20 nm, $E_0 = 100$ keV. (D.J. Johnson, unpublished.)

and it is possible that a visually acceptable image represents a false impression of the specimen. Any instrumental effect on an image must be treated with caution, and defocus must be chosen objectively.

Two examples will illustrate the effect of defocus on the image. Figure 2.3 shows a series of images of a transverse section of keratin microfibrils about 7.5 nm in diameter (Johnson and Crawford 1973). The 'in-focus' image (a) shows very little structure, but it can be seen that the image changes in appearance as defocus increases from (b) 600 nm to (c) 1000 nm to (d) 1600 nm. Image (b) can be interpreted on the basis of nine protofibrils of about 2 nm diameter in a ring surrounding two similar protofibrils, whereas image (d) represents another extreme of five protofibrils of about 3 nm diameter in a ring. Clearly a particular defocus is enhancing certain structural features and it is very difficult to say which image is the nearest to the actual substructure of keratin. It is possible, however, to explain the appearance of these images. For maximum phase contrast for a spacing r in the specimen the approximate value of defocus is:

$$\Delta f = r^2/2\lambda \tag{2.3}$$

for incident electrons of wavelength λ (Agar et al. 1974). Consequently a defocus of 600 nm will enhance spacings of 2.1 nm ($\lambda = 3.70$ pm for an incident electron energy of $E_0 = 100$ keV), whilst $\Delta f = 1600$ nm will enhance spacings of 3.4 nm, thus explaining the 'structural' interpretation of images (b) and (d) in Fig. 2.3. Depending on the choice of defocus certain spacings in the specimen will be enhanced and others will be suppressed or imaged with low contrast or even with reversed contrast. The conclusion is that none of the images in Fig. 2.3 can be believed unless further analysis is carried out (see Chapter 3).

A second example (Fig. 2.4) shows a focal series of negatively-stained groups of nine hexons from adenovirus. Image (a) corresponds to 'in-focus' ($\Delta f \simeq 50$ nm) and defocus increases in steps of 50 nm to (d) which is 200 nm underfocus. The dominant feature of all these images is the way the background structure from the supporting carbon film/negative stain changes with defocus, and little can be said about the morphology of the individual hexons, which are about 8 nm in diameter; it is evident that the internal features of the hexons are similar to the background structure. From these two examples it is clear that an objective choice of defocus is required, together with a method for measuring it; the appearance of the image cannot be safely used alone as a criterion. The effect of defocus in image analysis is examined in detail in

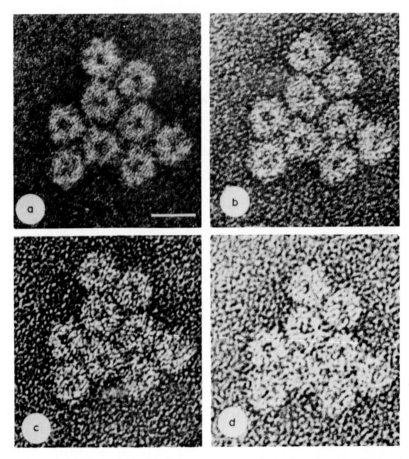

Fig. 2.4.   Bright-field focal series of groups of nine hexons from adenovirus, negatively-stained with sodium silicotungstate: (a) 'in-focus', (b) defocus $(\Delta f) = 100$ nm, (c) $\Delta f = 150$ nm and (d) $\Delta f = 200$ nm underfocus. Image bar $= 10$ nm, $E_0 = 80$ keV. (J.V. Heather, unpublished.)

Chapter 3. For the present it is useful to remember that Eq. (2.3) gives an approximate defocus for enhancing a spacing r in the specimen; also note that the 'in-focus' condition determined from observing the vanishing of granular structure (minimum contrast) in the carbon support film does not correspond to $\Delta f = 0$, but to an underfocus of approximately 50 nm (§ 3.3).

Other image defects, such as astigmatism, contamination of the objective aperture, and specimen drift, will show in the image as a streaking of the apparent grain structure of the supporting carbon film. This is a much more sensitive method of assessing the magnitude of these defects than the use of

Fresnel fringes around a hole in the carbon film, and, when used with optical analysis, gives a precise measure of them.

2.5 Contrast and resolution

Image contrast is defined in terms of the variation in optical density (image intensity) relative to the mean density of the image. In the bright-field microscopy of thin specimens the main contribution to the background is from the unscattered electron beam, whilst the contrast arises from differences in the electron scattering properties of the specimen. Because these differences are small in biological materials, staining techniques must generally be used to enhance them, otherwise the image will show no visible detail.

Contrast essentially arises from the *elastic* component of the scattered electron beam (Agar et al. 1974); these electrons are scattered by the nuclei of the atoms comprising the specimen through angles $\theta = 0.01-0.1$ rad and a large proportion of them are excluded by the objective aperture. Because elastic scattering increases with atomic number Z (the nuclear charge), the differences between unstained and stained parts of the biological specimen are enhanced by elastic scattering.

The second component of the scattered electron beam is called '*inelastic*'; it consists of electrons that lose energy in the specimen as a result of interacting with the atomic electrons. The inelastically scattered electrons transmitted by the specimen are mainly confined to a small angular range, $\theta = 0-0.002$ rad, compared to elastically scattered electrons. As a result of this small angular spread, most of the inelastically scattered electrons pass through the objective aperture and contribute to the image. Inelastic scattering is almost independent of Z so that it is most important for specimens of low Z or unstained specimens and is less important for specimens of high Z or stained specimens, where the elastic scattering dominates. Inelastic scattering is undesirable for several reasons: firstly this component is responsible for radiation damage in the specimen; secondly, because of the small scattering angles, it carries structural information only at low resolution; and thirdly, because the electrons lose energy in the specimen, they cannot be focused by the objective lens. Although an energy loss ΔE of about 25 eV for biological materials may seem to be insignificant in terms of an electron energy of 100 keV, the chromatic aberration constant C_c of the objective lens is large (1-2 mm). The resolution r_c of the inelastic image as limited by the chromatic defect is given by Cosslett (1956) as:

$$r_c = C_c \theta \Delta E / E_0 \tag{2.4}$$

where E_0 is the incident electron energy, ΔE is the energy loss and θ is the angle of scattering. For thin specimens (20 nm) the effective angle of scattering θ is small and the chromatic defect is not a serious limitation. However, as the specimen thickness increases (50 nm) θ and ΔE both increase as a result of multiple inelastic electron scattering in the specimen, and chromatic aberration finally limits the resolution of the image.

Perhaps a more satisfactory way of looking at the effect of inelastic scattering on the image is to note that an energy loss ΔE is equivalent to a defocus (Misell 1975):

$$\Delta f \simeq C_c \Delta E / E_0 \tag{2.5}$$

so that for $C_c = 2$ mm, $\Delta E = 25$ eV, $E_0 = 100,000$ eV, Eq. (2.5) gives $\Delta f = 500$ nm. Thus, assuming that focusing is based on optimising the high resolution elastic image, the inelastic image will be about 500 nm out of focus. Note also that the chromatic defect depends on the ratio $\Delta E / E_0$ in Eq. (2.4) (Cosslett 1969). One of the main objectives of image processing is the reduction of the amount of inelastic scattering in the reconstructed image because it contributes an image of poor quality and low contrast. There is no way of reducing the inelastic scattering in the specimen, although it is less significant in stained specimens than in unstained specimens (see § 2.5.2).

Resolution is difficult to define. It refers to the finest detail observed in the image that is consistent with the specimen structure; it does not refer to the smallest detail that can be observed in the image, which depends more on the intrinsic resolution of the electron microscope than on any property of the specimen. However, this is virtually impossible to quantify from the image alone and is indeed very difficult to assess for images of non-periodic structures, even with the aid of optical diffractometry. If the specimen has a repeating structure, however, optical diffractometry can provide a clear separation of structural and 'non-structural' (e.g. from the carbon support film) information. Image analysis and image processing applied to structures showing symmetry (translational, helical or rotational) enable a number to be assigned to the resolution of the image and this number refers to the specimen rather than to microscope performance. For example, from the images shown in Fig. 2.4, the smallest detail that can be resolved is about 0.5 nm, but rotational filtering (see § 4.1) on these groups of nine hexons ($n = 3$) shows that the actual hexon structure within the group of nine is preserved only to a resolution of about 5 nm.

2.5.1 Contrast mechanisms

One of the commonest ways of obtaining contrast in the image is by defocusing the objective lens until maximum contrast is achieved (Agar et al. 1974). If the appearance of an image is sensitive to these changes in focus then the contrast is almost certainly arising from phase contrast: that is, contrast resulting from the interference of the unscattered and the elastically scattered electrons, which are shifted in phase by the effect of the defocus Δf. As in phase contrast light microscopy, the ideal phase shift is $\pi/2$ or 1/4 of a wavelength, but in electron microscopy this can be achieved only approximately. The choice of Δf depends on the size of the specimen spacings to be imaged, as shown by Eq. (2.3). Thus in a negatively-stained specimen containing structures with a spacing of 2 nm, the defocus giving maximum phase contrast is 540 nm (Erickson and Klug 1971); of course, spacings above and below 2 nm will not be imaged with maximum phase contrast.

The structural information in biological specimens is rarely below 2 nm, and consequently an 'in-focus' image of such a specimen will show very low phase contrast. The main contribution to image contrast for 'in-focus' settings is from scattering contrast; that is, the differential scattering of electrons outside the objective aperture from regions of the specimen with different density. Scattering contrast depends on the angular distribution of the scattered electrons and the size of the objective aperture. It increases with increasing specimen thickness (the average value of θ increases, see § 2.5.2) and decreasing size of the objective aperture, because the differential effect of scattering is enhanced. (Phase contrast shows very little dependence on objective aperture size and may actually decrease with increasing specimen thickness.) Most observations on thin sections are done under scattering contrast conditions, with images recorded just underfocus. In a typical section showing structural detail down to 5 nm, phase contrast would only be effective for a defocus of about 3.4 μm. The reason that phase contrast is preferred for imaging *very* thin specimens is because the maximum contrast to be expected is a factor of four or five times greater than the contrast which can be obtained by scattering contrast alone.

In practice, most images consist of contributions from both phase and scattering contrast, but experimental evidence indicates that images taken with an appropriate choice of defocus for phase contrast show only a small contribution from scattering contrast. In this book preference is given to the phase contrast mechanism of image formation because the effects of defocus can be described quantitatively, and in addition, more information is available about the specimen structure from such an image (see Chapter 3); in this approach the image is considered to be formed by the interference between the electrons transmitted by the specimen.

The wave-optical theory of image formation cannot only be used to explain phase contrast images but is also flexible enough to include the effects of scattering contrast. The reverse is not true: scattering contrast cannot be modified to include phase contrast because it is essentially a particle description of image formation; this type of model cannot, for example, be used to explain electron diffraction from crystalline specimens. The image analysis and image processing methods given in Chapters 3 and 4 are based on wave-optics and the interpretation of the results will evidently be influenced by how correct this model is for image formation. However, without such a model for the relationship between the specimen structure and its image, it is not possible to make any interpretation of the image. Early papers on image reconstruction assumed that the images of negatively-stained specimens representea a two-dimensional projection of three-dimensional density variations in the stain, and an essentially scattering contrast mechanism for image formation; this is not necessarily incorrect, and may in fact give satisfactory reconstructions at low resolution (Frank 1973). However, it is now more usual to accept the phase contrast interpretation of images of negatively-stained specimens, because it does take account explicitly of the effect of objective lens defocus on the image.

The preceding discussion refers to bright-field microscopy when the objective aperture is centred so that the unscattered electron beam is transmitted; this is the most common type of microscopy used in biology. In dark-field microscopy the unscattered electrons are prevented from contributing to the image; this is achieved by either tilting the incident electron beam or by displacing the objective aperture so that the intense unscattered radiation is not transmitted (Kleinschmidt 1971). In dark-field microscopy the phase contrast effects are small compared to bright-field microscopy and the image seems to be less dependent on defocus. However, the scattering contrast may be increased because the image no longer contains a large background contribution from the unscattered component as in bright-field microscopy. Thus if the contrast of a biological specimen is low in bright-field microscopy then it may be advantageous to change to dark-field microscopy using, for example, the dark-field tilt controls on many modern microscopes (see § 5.4).

2.5.2   *Elastic and inelastic scattering*

The essential difference between elastic and inelastic scattering is that the inelastically scattered electrons emerge from the specimen with a lower energy than the electrons in the incident beam. Thus, if the transmitted electron beam is considered in terms of energy loss ΔE, the unscattered and elastically scattered electrons constitute the sharp peak centred at $\Delta E = 0$, whilst the inelastically scattered electrons are spread out into an energy loss spectrum as shown in Fig. 2.5. For a thin specimen the energy loss spectrum

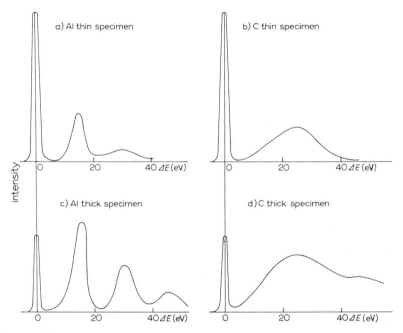

Fig. 2.5.   The energy loss spectra of aluminium and carbon for a thin and a thick specimen. ΔE is the energy loss.

consists of a single peak, which may be narrow, as for aluminium (a few electron volts wide; Fig. 2.5a), or broad, as for carbon and biological materials (20 eV wide; Fig. 2.5b). When the objective lens is set to focus the elastic component ($\Delta E = 0$), the inelastic image will comprise a series of images becoming increasingly out of focus as the energy loss increases ($\Delta f \simeq C_c \Delta E / E_0$). As the specimen thickness increases, electrons are inelastically scattered more than once and this leads to multiple peaks in the energy loss spectrum

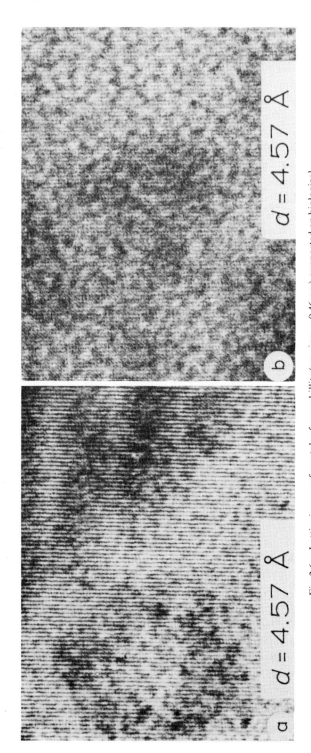

Fig. 2.6. Lattice images of crystals of pyrophillite (spacing = 0.46 nm), supported on biological sections: (a) thin section (30 nm), (b) thick section (300 nm). $E_0 = 75$ keV. (From Nagata and Hama 1971.)

and an overall increase in the average energy loss (see Fig. 2.5c and d). Consequently, the chromatic defect or blurring of the inelastic image becomes larger as the specimen thickness increases. Also note that as the specimen thickness increases the unscattered component diminishes and a certain fraction of the elastically scattered electrons is subsequently inelastically scattered and contribute to the blurred image.

In thick specimens (0.2–1 μm) nearly all the electrons will be inelastically scattered at least once, and the image resolution will be limited by the chromatic aberration of the objective lens, which cannot be corrected by image processing. Thus, whereas for a thin specimen the inelastic component may form an undesirable background on which the high resolution elastic image is superimposed, for a thick specimen this high resolution detail will be blurred by chromatic aberration leading to a drastic reduction in image contrast. This effect is shown in Fig. 2.6 where a thin crystal of pyrophillite is supported on a (a) thin (30 nm) and (b) thick (300 nm) section of biological material (Nagata and Hama 1971). The 0.46 nm lattice spacing is clearly visible in Fig. 2.6a but its visibility (contrast) is severely reduced when the biological specimen is thick (Fig. 2.6b).

The main consequence of inelastic scattering then, is a loss in image contrast, and so its removal either instrumentally or by image processing is desirable. Instrumentally an electron spectrometer may be used to distinguish unambiguously between elastically and inelastically scattered electrons, and this is the method used in the energy-selecting electron microscope (see § 5.2) to produce energy-filtered images.

Elastically and inelastically scattered electrons can also be discriminated on the basis of their angular distributions. For a thin specimen the inelastic scattering is confined to small angles of scattering (0.002 rad), whilst the elastic scattering is predominant at larger angles. A comparison of the angular distributions for scattering in carbon (corresponding approximately to unstained biological sections) is shown in Fig. 2.7. Note that the intensity axis is a logarithmic scale. For $E_0 = 100$ keV and a thin specimen (20 nm) the inelastic scattering (I) exceeds the elastic scattering (E) for angles less than 0.01 rad. However, as the specimen thickness increases, or the incident electron energy is decreased, the inelastic scattering exceeds the elastic scattering over a large range of angles. This is a result of combined elastic–inelastic scattering which broadens the inelastic angular distribution (Fig. 2.7). Since the normal objective aperture corresponds to a semi-angle α_{obj} in the range 0.01–0.02 rad (α_{obj} = objective aperture diameter/2 × focal length of the objective lens = $d_{obj}/2f_{obj}$; Agar et al. 1974), an estimate can

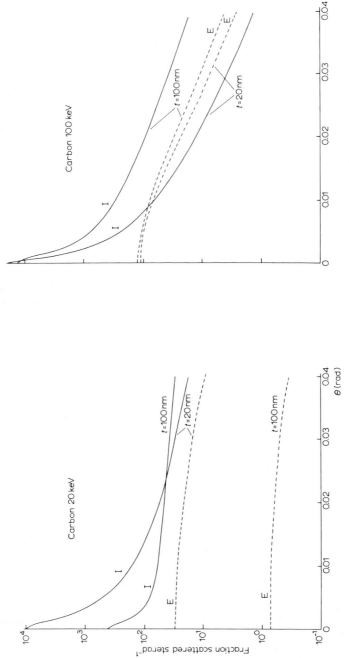

Fig. 2.7.   Electron scattering in carbon: a comparison of the angular distributions for elastic (E) and inelastic (I) electron scattering for incident electrons of energy 20 keV and 100 keV with specimens of thickness 20 nm and 100 nm.

be made of the relative contributions of elastic and inelastic scattering to the image of an unstained section; for $\alpha_{obj} = 0.01$ rad the ratio of inelastic to elastic scattering I/E, is as high as 10 : 1 for a specimen 20 nm thick observed at 100 keV (Burge 1973). Of course, this ratio is not so unfavourable for stained specimens when the elastic scattering becomes relatively more important, nor has account been taken of the fact that the elastic image may be enhanced by phase contrast as a result of the interference of the elastically scattered and the unscattered electrons.

It can be concluded that an approximate separation of the elastic and inelastic scattering can be made on the basis of the respective angular distributions for a thin specimen. This is the principle used in the scanning transmission electron microscope (STEM) where an annular detector is used to collect the elastically scattered electrons, allowing most of the inelastic (and unscattered) electrons to pass through the centre of the detector (see § 5.3). However, such a discrimination is not possible for thick unstained specimens.

Although elastic scattering does increase relative to the inelastic scattering as the average atomic number Z of the specimen increases, as soon as the specimen thickness reaches the mean free path for inelastic scattering (about 50 nm), there is an increasing probability that elastically scattered electrons will be inelastically scattered as well. Thus generally, as the specimen thickness increases, the ratio I/E increases, irrespective of Z, as shown in Fig. 2.8. So inelastic scattering cannot necessarily be disregarded even in stained specimens or specimens containing high Z material.

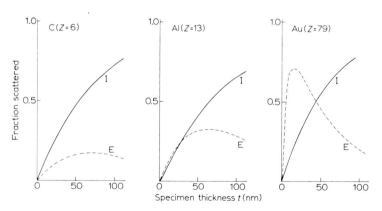

Fig. 2.8.   The relative proportions of elastic (E) and inelastic (I) scattering in different materials as the specimen thickness increases (no objective aperture). $E_0 = 100$ keV.

The recommendation is to use specimens as thin as possible so as to reduce the detrimental effect of inelastic scattering on image contrast and image resolution. This reinforces the reasons given in § 2.3 for using thin specimens whenever possible. However, if thick specimens are unavoidable, image processing can reduce the adverse effects of inelastic scattering provided that the specimen has some symmetry (see Chapters 3 and 4). Instrumentally the detrimental effects of inelastic scattering may be reduced by using an energy-selecting electron microscope or a STEM (see § 5.2 and § 5.3). In certain specialised high-resolution applications, such as resolving single heavy atoms in a biological macromolecule, the image subtraction procedure detailed in § 6.2 may be used; this is based on the differential behaviour of the elastic and inelastic images as the objective lens defocus is varied (Misell and Burge 1975).

References

Agar, A.W., R.H. Alderson and D. Chescoe (1974), Principles and practice of electron microscope operation, in: Practical methods in electron microscopy, Vol. 2, A.M. Glauert, ed. (North-Holland, Amsterdam).

Aebi, U., P.R. Smith, J. Dubochet, C. Henry and E. Kellenberger (1973), A study of the structure of the T-layer of *Bacillus brevis*, J. Supramol. Struct. *1*, 498–522.

Beer, M., J. Frank, K.-J. Hanszen, E. Kellenberger and R.C. Williams (1975), The possibilities and prospects of obtaining high-resolution information (below 30 Å) on biological material using the electron microscope, Biophys. J. *7*, 211–238.

Beeston, B.E.P. (1972), An introduction to electron diffraction, in: Practical methods in electron microscopy, Vol. 1, part 2, A.M. Glauert, ed. (North-Holland, Amsterdam), pp. 193–323.

Burge, R.E. (1973), Mechanisms of contrast and image formation of biological specimens in the transmission electron microscope, J. Microsc. (Oxford) *98*, 251–285.

Cosslett, V.E. (1956), Specimen thickness and image resolution in electron microscopy, Brit. J. Appl. Phys. *7*, 10–12.

Cosslett, V.E. (1969), Energy loss and chromatic aberration in electron microscopy, Z. Angew. Phys. *27*, 138–141.

Crowther, R.A. (1971), Procedures for three-dimensional reconstruction of spherical viruses by Fourier synthesis from electron micrographs, Phil. Trans. Roy. Soc. Lond. B *261*, 221–230.

Crowther, R.A. and A. Klug (1975), Structural analysis of macromolecular assemblies by image reconstruction from electron micrographs, Ann. Rev. Biochem. *44*, 161–182.

Crowther, R.A., D.J. DeRosier and A. Klug (1970), The reconstruction of a three-dimensional structure from projections and its application to electron microscopy, Proc. Roy. Soc. Lond. A *317*, 319–340.

Erickson, H.P. and A. Klug (1971), Measurement and compensation of defocusing and aberrations by Fourier processing of electron micrographs, Phil. Trans. Roy. Soc. Lond. B *261*, 105–118.

Frank, J. (1973), Computer processing of electron micrographs, in: Advanced techniques in biological electron microscopy, J.K. Koehler, ed. (Springer-Verlag, Berlin), pp. 215–274.

Glaeser, R.M. (1975), Radiation damage and biological electron microscopy, in: Physical aspects of electron microscopy, B.M. Siegel and D.R. Beaman, eds. (John Wiley and Sons, New York), pp. 205–229.

Horne, R.W. (1973), Contrast and resolution from biological objects examined in the electron microscope with particular reference to negatively stained specimens, J. Microsc. (Oxford) *98*, 286–298.

Johansen, B.V. (1976), Bright-field electron microscopy of biological specimens. V. A low dose pre-irradiation procedure reducing beam damage, Micron *7*, 145–156.

Johnson, D.J. and D. Crawford (1973), Defocusing phase contrast effects in electron microscopy, J. Microsc. (Oxford) *98*, 313–324.

Kleinschmidt, A.K. (1971), Electron microscopic studies of macromolecules without appositional contrast, Phil. Trans. Roy. Soc. Lond. B *261*, 143–149.

Klug, A. (1971), Applications of image analysis techniques in electron microscopy. III. Optical diffraction and filtering and three-dimensional reconstruction from electron micrographs, Phil. Trans. Roy. Soc. Lond. B *261*, 173–179.

Klug, A. and D.J. DeRosier (1966), Optical filtering of electron micrographs: reconstruction of one-sided images, Nature *212*, 29–32.

Misell, D.L. (1975), The resolution and contrast in biological sections determined by inelastic and elastic scattering, in: Physical aspects of electron microscopy, B.M. Siegel and D.R. Beaman, eds. (John Wiley and Sons, New York), pp. 63–79.

Misell, D.L. and R.E. Burge (1975), Contrast enhancement in biological electron microscopy using two micrographs, J. Microsc. (Oxford) *103*, 195–202.

Nagata, F. and K. Hama (1971), Chromatic aberration on electron microscope image of biological sectioned specimen, J. Electron Microsc. *20*, 172–176.

Stenn, K. and G.F. Bahr (1970), Specimen damage caused by the beam of the transmission electron microscope, a correlative reconsideration, J. Ultrastruct. Res. *31*, 526–550.

Unwin, P.N.T. (1974), Electron microscopy of the stacked disc aggregate of tobacco mosaic virus protein. II. The influence of electron irradiation on the stain distribution, J. Mol. Biol. *87*, 657–670.

Unwin, P.N.T. (1975), Beef liver catalase structure: interpretation of electron micrographs, J. Mol. Biol. *98*, 235–242.

Unwin, P.N.T. and R. Henderson (1975), Molecular structure determination by electron microscopy of unstained crystalline specimens, J. Mol. Biol. *94*, 425–440.

Williams, R.C. and H.W. Fisher (1970), Electron microscopy of tobacco mosaic virus under conditions of minimal beam exposure, J. Mol. Biol. *52*, 121–123.

Chapter 3

Image analysis

Confirmation or the extension of the information obtained by a visual assessment of an electron micrograph is obtained by *image analysis*. The use of either optical analysis (§ 3.3) or *computer analysis* (§ 3.4) enables a quantitative assessment of image defects, such as astigmatism and specimen drift, and a determination of image defocus to be made (§ 3.3.2 and § 3.3.4). For images obtained from ordered specimens, image analysis based on the *Fourier transform* of the image gives an indication of how well the structure is preserved, in addition to the value of the average spacing between structural units in the lattice (§ 3.3.3).

As a prerequisite to this type of image analysis, it is necessary to understand the mathematical basis of the *Fourier analysis* of images (§ 3.1), and to establish a mathematical relationship between the *optical density variations* in the image or *image intensity* and the specimen structure (§ 3.2). Provided that the specimen is thin (§ 3.2.1), the relationship between the image intensity and specimen structure is *linear* (§ 3.2). Then the *image transform* is a simple multiplication of a microscope-dependent quantity, called the *transfer function*, and the Fourier transform of the specimen structure. The image transform can then be used to assess both *image quality* (§ 3.3.2) and image resolution (§ 3.3.3).

More advanced procedures of image analysis are described in § 3.6 and § 3.7. The analysis of transforms obtained from images of helical particles is described in § 3.6 and § 3.7.3. *Indexation* of the transform obtained from the image of an ordered structure corresponds to assigning the positions of maximum intensity in the transform (*diffraction maxima*) to a single reciprocal lattice (§ 3.7.1). For specimens consisting of two superimposed layers, indexation corresponds to choosing two similar reciprocal lattices rotated with respect to each other (§ 3.7.2). Indexation is an essential prerequisite to the processing and enhancement of images obtained from ordered structures (§ 4.2 and § 4.3).

3.1   The Fourier transform

Images are analysed by examination of the *Fourier transform* of the electron micrograph, obtained optically (§ 3.3) or numerically (§ 3.4). This Fourier transform is an alternative way of assessing images and does not exclude the normal visual assessment; it corresponds to a *frequency analysis* of the image. The image is considered to be the sum of a set of sinusoidal waves of varying frequency; the higher the frequency of the sine wave, the finer the detail it refers to. Mathematically this is expressed by writing the variation of the image intensity (optical density) $j(x)$ with position x in the image as:

$$j(x) = \sum_h J_h \sin\left(\frac{2\pi h x}{l}\right) \tag{3.1}$$

assuming that $j(x)$ has a periodicity of l; J_h are referred to as Fourier coefficients and they represent the weighting factor given to a particular sine wave of frequency $v = h/l$. A small value of h implies a slowly varying sine wave, which accounts for coarse (low resolution) detail in the image, whilst a large value of h indicates fine (high resolution) detail. Thus there is an inverse relation between the image and its Fourier transform: small spacings r correspond to high frequencies v. Whereas resolution may be difficult to assess visually from an image, its transform indicates the maximum frequency contained in the image, and by implication the resolution of the image. The unit of measurement in the micrograph is length, so the unit of v is reciprocal length, usually nm^{-1}. Any measurement of v can be related to a spacing r in the image by $r = 1/v$. Note that although it is the image that it is analysed, it is spacings in the specimen that are required. In this book r and v will normally refer to the specimen; the corresponding quantities are Mr and v/M in the image taken at magnification M.

Equation (3.1) has been simplified because the coefficients of J_h are generally complex numbers, with a real and an imaginary part. This means that J_h not only has an *amplitude* (strengths of the sine waves) but also a *phase angle* (positions of the sine waves) associated with it, so that the general form of Eq. (3.1) is:

$$j(x) = \sum_h J_h \exp\left(\frac{2\pi i h x}{l}\right)$$

$$= \sum_h J_h \left[\cos\left(\frac{2\pi h x}{l}\right) + i \sin\left(\frac{2\pi h x}{l}\right)\right] \tag{3.2}$$

and the Fourier coefficients J_h are calculated from the image using the inverse relation:

$$J_h = \frac{1}{l} \int_0^l j(x) \exp\left(\frac{-2\pi i h x}{l}\right) dx \tag{3.3}$$

The important thing to remember in Fourier analysis is that the coefficients J_h include two pieces of information, firstly the magnitude or amplitude of J_h, $|J_h|$, and secondly its phase ω_h: $J_h = |J_h| \exp(i\omega_h)$. It is difficult to attach a precise meaning to ω_h (as is possible for $|J_h|$, the relative weighting of the sine and cosine waves in Eq. (3.2)), but ω_h is an essential part of the Fourier coefficient determining the positional information (x) in the image $j(x)$. For example, if only the *modulus* (magnitude) of J_h is measured, then it is impossible to reconstitute the image $j(x)$ from Eq. (3.2). This is very similar to the phase problem in X-ray crystallography (Sherwood 1976), where there is a Fourier relationship between the electron density distribution in the crystal $\rho(x)$ and the diffraction pattern F_h (equivalent to J_h used above); only the intensities of the diffraction pattern $|F_h|^2$ can be measured and it is impossible to determine the structure of the crystal $\rho(x)$ without some knowledge of the phases of F_h, ω_h. It is also true that with relatively poor amplitudes $|F_h|$ but good phases ω_h, a structure can be determined. Essentially the phases contain information on the *positions* of the density variations $\rho(x)$ while the amplitudes determine their *magnitude*. This emphasises the importance of knowing both the amplitude and phase of J_h in the Fourier series in Eq. (3.2). Without the information on ω_h, optical density variations in $j(x)$ occur in incorrect positions and a reconstructed image would be impossible to interpret. Fortunately in electron microscopy there is no analogue of the X-ray phase problem, provided the specimen is thin, and so the transform of an image can be used to determine the specimen structure (§ 3.2).

For completeness, Eq. (3.2) should be written in a general form for a two-dimensional image which does not have a period l. The Fourier coefficients J_h then become a continuous function of frequency in two dimensions and the summation in Eq. (3.2) becomes an integration over all frequencies:

$$j(x, y) = \int\limits_{-\infty}^{+\infty} \int\limits_{-\infty}^{+\infty} J(v_x, v_y) \exp \left[2\pi i(v_x x + v_y y) \right] dv_x \, dv_y \qquad (3.4)$$

The transform still can be interpreted in a similar way to the Fourier series, where $J(v_x, v_y)$ represents the weighting given to frequencies v_x and v_y in the x and y directions. It is also convenient to use a shorthand notation for a two-dimensional quantity (vector): r is used to represent a coordinate (x, y) in the image, and v represents a two-dimensional frequency (v_x, v_y) in the transform. On this basis, a Fourier transform is defined as:

$$J(v) = \int j(r) \exp \left(-2\pi i v \cdot r \right) dr \qquad (3.5)$$

and its inverse transform as:

$$j(r) = \int J(v) \exp \left(2\pi i v \cdot r \right) dv \qquad (3.6)$$

These equations are the mathematical basis of image analysis.

Thus the Fourier transform of an image is a frequency representation of the image, which will often be more informative than purely visual assessment. The transform coefficients $J(v)$ have two components, an amplitude $|J|$ and a phase ω. If the transform is determined optically, the photographic film used to record the transform records only intensity $|J|^2$ (§ 3.3). A *numerical transform* can be determined after converting optical density variations in the micrograph to numbers using a scanning densitometer, and gives both $|J|$ and ω (§ 3.4).

In order to illustrate the advantages of the Fourier transform in the analysis of images, some examples are given below.

3.1.1   Detection of image defects and amount of defocus

It is very difficult to assess quantitatively the image defects and the amount of defocus by direct inspection of an image of a biological specimen. The objective lens defects (spherical aberration and astigmatism) and defocus cause distortions in the image; the transfer of information from the specimen to the image is then imperfect. The easiest way to visualise these distortions is to consider how the microscope affects particular frequencies (spacings) in the specimen, in a similar way to the assessment of a Hi-Fi system in terms of its frequency response. Thus, if the microscope were perfect, all frequencies present in the specimen would be present in the image with their correct weighting. But lens defects and defocus cause certain frequencies to be imaged with low contrast or even reversed contrast. An examination of the transform of an image immediately gives some indication of how the frequency spectrum has been affected and, in particular, the shape of the transform profile can be used to determine the magnitude of the image defects and a value for the image defocus (see § 3.3).

3.1.2 Analysis of a regular array of subunits

If the specimen contains a regular array of subunits, direct examination of the image will usually only confirm that the specimen is periodic, but it will not indicate how regular the arrangement is, nor at what resolution the internal features of the subunits are preserved. Of course, the tedious procedure of measuring spacings between subunits in the image can give some indication of the order in the specimen, but a transform gives the answer immediately.

The transform of a perfect (undistorted) two-dimensional array of subunits consists of a set of sharp spots whose spacing depends on the original separation of the subunits (e.g. § 3.3; Figs. 3.6 and 3.7). Mathematically, a one-dimensional crystal $f(x)$ can be described as a sum of subunits (motifs) $g(x)$ with a separation of a:

$$f(x) = g(x) + g(x - a) + g(x - 2a) + \ldots$$

or more formally:

$$f(x) = \sum_n g(x)\delta(x - na) \tag{3.7}$$

where the function $\delta(x - na)$ represents the lattice and $g(x)$ represents the *motif* or basic shape of the subunit. The delta function $\delta(x - na)$ is zero unless x is some integral multiple n of the *lattice constant a* and unity if $x = na$. The Fourier transform of Eq. (3.7) is:

$$F(v_x) = G(v_x) \sum_h \delta(v_x - h/a) \tag{3.8}$$

so that the Fourier transform is only non-zero if v_x is some integral multiple h of $1/a$. Thus the transform consists of an array of intensity maxima spaced at h/a, whilst the intensities of the diffraction maxima give information on $G(h/a)$, the Fourier transform of the motif $g(x)$. Again Eq. (3.8) emphasizes the reciprocal relation between the image and its transform: a distance 'a' in $f(x)$ becomes a separation $1/a$ in the transform $F(v_x)$.

Distortions of a regular structure in the image show up as blurring of the diffraction maxima (§ 3.3.3; Fig. 3.17) and a measurement of the broadening of the maxima gives a measure of the variations in a of the original lattice. In addition, the highest order diffraction maximum (maximum value of h) observed gives the maximum frequency h_{max}/a in the transform and by implication the resolution of the subunit structure a/h_{max}. For example, if a lattice with a spacing of 10 nm gives a diffraction pattern with five orders measured from the central maximum, then $h = 5$, and, with some certainty, it can be stated that the structure of the subunits is preserved to a resolution of $10/5 = 2$ nm. The more diffraction orders observed, the higher the resolution achieved, measured with respect to the lattice spacing a.

Again Eqs. (3.7) and (3.8) must be extended for application to two-dimensional lattices. The lattice will now be characterised by two lattice vectors, a and b corresponding to lattice spacings $a = |a|$ and $b = |b|$. a and b must be expressed as vectors because they have magnitude and direction; only when a is along the x-axis and b along the y-axis can they be considered as simple quantities a and b. Thus in two-dimensions Eq. (3.7) becomes:

$$f(r) = \sum_n \sum_m g(r)\delta(r - na - mb) \tag{3.9}$$

and its Fourier transform is:

$$F(v) = G(v) \sum_h \sum_k \delta(v - ha^* - kb^*) \tag{3.10}$$

where $a^* = 1/a$ and $b^* = 1/b$ are referred to as the *reciprocal lattice vectors*, and these determine the positions of the intensity maxima in the *reciprocal lattice*. Note that if the two principal axes of the transform (i.e. along a^* and b^*) are inclined at an angle α, the lattice spacings a and b are given by:

$$a = \frac{1}{a^* \sin \alpha} \,, \quad b = \frac{1}{b^* \sin \alpha} \tag{3.11}$$

and not simply by $1/a^*$ and $1/b^*$. So in hexagonal lattices, which occur in a number of membrane systems (e.g. in gap junctions and the outer envelopes of *E. coli* and other bacteria), $a^* = b^*$ and $\alpha = 60°$, so that the actual hexagonal lattice spacing is given by $a = 1/a^* \sin 60° = \sqrt{4/3} \times 1/a^*$ and *not* by $1/a^*$. Incorrect lattice spacings will be calculated if the spacing between diffraction maxima is used directly to calculate a and b, without taking account of the angle between the principal reciprocal lattice vectors.

The analysis here is restricted to two dimensions, because an electron micrograph corresponds to a two-dimensional projection. But remember that the specimen is a three-dimensional structure and that these two-dimensional projections may include information on all three lattice vectors, *a*, *b* and *c* (see Chapter 7).

3.1.3 Separation of information from overlapping layers

Many biological structures are tubular when intact, but after drying flatten on the carbon support film to become two nominally identical layers superimposed on each other, but inclined at an angle α. Examples include microtubules, whole bacterial membranes and T4 phage heads. Multilayer assemblies such as acetylcholine receptor membranes (Ross et al. 1977) and ribosome crystals (Unwin and Taddei 1977) also occur. In the image this gives a complex moiré pattern from two or more slightly misaligned and superimposed lattices. Direct examination of the image is not very informative; in fact it may be difficult to tell the difference between poor specimen preparation and a good image which is merely confused by the superposition of layers. The transform clarifies this situation immediately, because the respective transforms of the layers will be rotated with respect to each other. Thus, for example, diffraction spots from two layers will occur in pairs and it is a relatively easy task to check that these pairs of maxima correspond to two lattices inclined at an angle α (§ 3.7). An assessment can then be made of the quality of the image. Although these multilayered specimens can sometimes be treated biochemically to give single layers, the intact double-layered structure is often better preserved. This will be evident from the transform, by examining the extent and width of the diffraction maxima.

3.1.4 Analysis of helices

Many biological complexes, such as some nucleoproteins and bacterial flagella, aggregate into structures with helical symmetry. This symmetry may be directly evident in the image, especially if the specimen is metal shadowed, but the Fourier transform of the image will give a characteristic helical diffraction pattern from which the parameters of the helix may be determined (i.e. the spacings of the subunits along the helix, the screw angle of the helix, and in complex helical structures, the number of basic helices involved (§ 3.6 and § 3.7)).

3.1.5   Analysis of particles with rotational symmetry

It will be evident that Fourier analysis of electron images is most fruitful when the specimen exhibits periodicity, because the transform shows this symmetry clearly. The other type of symmetry mentioned previously, rotational symmetry (Chapter 1), is not so evident in the optical transform, because rotationally symmetrical particles do not give discrete transforms. Computer analysis of numerical transforms is more useful (Crowther and Amos 1971) because by examining the phase information, some indication of the type of symmetry can be obtained.

For a non-periodic specimen the transform is not discrete but continuous, and little can be said about the resolution of the specimen from the transform alone, because it is usually impossible to separate the transform of the specimen from that of the carbon support film, which is also non-periodic. For non-periodic specimens the main application of Fourier analysis is the assessment of image quality rather than the properties of the specimen (§ 3.3).

3.2   Theory of image formation for a thin specimen

The first step in image analysis is to derive a relationship between the image and the specimen structure. A physical description of the interaction of the incident electron beam with a thin specimen in a transmission electron microscope is provided by the *phase object approximation* (Cowley and Moodie 1957). In this model the *potential distribution* in the specimen $V(r, z)$ affects the incident electron wave just as the refractive index affects a light wave in optics; r refers to the (x, y) coordinate in the plane of the specimen and the z-direction corresponds to the direction of the incident electron beam. The purpose of electron microscopy is to determine $V(r, z)$ or its two-dimensional projection, from measurements of the image intensity $j(r)$. In the phase object approximation the electron wavefunction after transmission through the specimen is:

$$\psi_0(r) = \exp\left[-i\eta(r)\right] \qquad (3.12)$$

corresponding to a phase shift of $\eta(r)$ radians, relative to the incident electron wave, proportional to the projection of the potential distribution $V(r, z)$ onto a two-dimensional plane $r = (x, y)$ in the specimen (that is, an integration in the z-direction). The i which precedes η in Eq. (3.12) is used mathematically to denote a phase shift in the exponential; $\exp(-\eta)$ would express an amplitude attenuation. $\operatorname{Exp}(-i\eta) = \cos \eta - i \sin \eta$ is a general way of writing the phase effect of a phase shift η; $i = \sqrt{-1}$. For a specimen of thickness t:

$$\eta(r) = \sigma \int_0^t V(r, z)\, dz \tag{3.13}$$

σ is a constant $= 2\pi me\lambda/h^2 = \pi/\lambda E_0$ for electrons of wavelength λ, energy E_0, mass m and charge e. In order to arrive at this simple result *coherent illumination* of the specimen is assumed; that is, that the incident electron beam is collimated and has zero energy spread. The effects of limited source coherence on the image and the Fourier transform of the image will be examined qualitatively in § 3.3. Also the effect of specimen thickness on the electron wavefunction ψ_0 has been neglected; in simple terms, an electron scattered at the top of the specimen will have a different phase from an electron scattered at the bottom of the specimen, corresponding to a defocus difference $\Delta f = t$. The thickness effect and the limitations it imposes on the interpretation of electron images will be examined in § 7.3. Finally the phase object approximation takes account of elastic scattering only and does not include the effect of inelastic electron scattering. The image formed by the inelastically scattered electrons is of intrinsically low resolution. Thus the final (elastic + inelastic) image can be considered as a superposition of a high resolution elastic image and a low resolution inelastic image. For the present it will be assumed that the inelastic scattering forms an undesirable low resolution background. The indirect effect of inelastic scattering in depleting the elastic scattering can be taken into account by including an absorption term in Eq. (3.12), $\exp[-\varepsilon(r)]$:

$$\psi_0(r) = \exp[-i\eta(r)] \exp[-\varepsilon(r)] \tag{3.14}$$

The first two approximations made in order to obtain a simple relationship between $\eta(r)$ and the image intensity $j(r)$ are:

(i) The effect of inelastic scattering on ψ_0 is neglected;

(ii) The phase shift $\eta(r)$ is small enough so that $\exp\left[-i\eta(r)\right]$ in Eq. (3.12) can be replaced by $1 - i\eta(r)$, neglecting terms containing η^2 and higher powers of η.

The limitations that this *weak phase approximation* places on specimen thickness are considered later in this section.

The *diffracted* (scattered) wave $S(v)$ corresponding to v is calculated from the two-dimensional Fourier transform of the object wavefunction $\psi_0(r)$ (Lenz 1971):

$$S(v) = \int \psi_0(r) \exp\left(-2\pi i v \cdot r\right) dr \qquad (3.15)$$

The use of the spatial frequency v is preferred to using the angle of scattering θ because $1/v$ can be interpreted directly in terms of resolution independently of electron wavelength λ; θ and v are related by $\theta = \lambda v$ for small angles of scattering so that the magnitude of v, $|v| = \theta/\lambda$. (Electrons scattered through an angle θ contribute to the image a component with frequency v.) Experimentally only the intensity $|S|^2$ of the scattered wave can be measured. The diffracted or scattered intensity is normally observed in the back focal plane (diffraction plane) of the objective lens. In the weak phase approximation:

$$S(v) = \delta(v) - iA(v) \qquad (3.16)$$

where $A(v)$ is the two-dimensional Fourier transform of $\eta(r)$; the Dirac delta function, $\delta(v)$, represents the intense unscattered beam at zero angle of scattering. The electron diffraction intensities for $v \neq 0$ are given by:

$$I(v) = |S(v)|^2 = |A(v)|^2 = A_r^2(v) + A_i^2(v) \qquad (3.17)$$

where A_r and A_i are the real and imaginary parts of the complex function $A = A_r + iA_i$. Thus from electron diffraction only the amplitude (magnitude) of A, $|A|$, can be determined. It is not possible to directly determine η from $|A|$, because the phase information concerning A is lost in the diffraction pattern.

It is conventional to use the diffraction plane (objective aperture plane) to describe the effects of an objective aperture, lens defects and defocusing on the object wavefunction (Fig. 3.1). The effect of spherical aberration

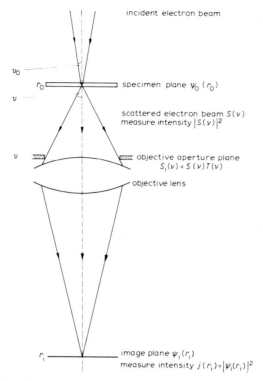

incident electron beam

v_0

r_0 specimen plane $\psi_0\,(r_0)$

v

scattered electron beam $S(v)$
measure intensity $|S(v)|^2$

v objective aperture plane
$S_i(v) = S(v)T(v)$

objective lens

r_i image plane $\psi_i(r_i)$
measure intensity $j(r_i) = |\psi_i(r_i)|^2$

Fig. 3.1.   The various planes in a conventional transmission electron microscope.

(coefficient C_s), defocusing (Δf) and axial astigmatism (coefficient C_a) is effected by multiplying $S(v)$ by the *transfer function* (Lenz 1971):

$$T(v) = \exp\left[-i\gamma(v)\right]B(v) \tag{3.18}$$

where the phase shift $-\gamma(v)$, introduced by the lens aberrations, is given by:

$$\gamma(v) = \frac{2\pi}{\lambda}\left[\frac{C_s v^4 \lambda^4}{4} - \frac{\Delta f v^2 \lambda^2}{2} - \frac{C_a \lambda^2}{2}(v_x^2 - v_y^2)\right] \tag{3.19}$$

and $B(v)$ is the objective aperture function (defined below). The convention used here is that positive $\Delta f(\Delta f > 0)$ corresponds to underfocus of the objective lens. Overfocus ($\Delta f < 0$) leads to an increase in the phase shift $-\gamma$, as a result of the addition of the spherical aberration and defocus terms. The ideal transfer function would correspond to one that introduces no phase shift into S, that is, $\gamma = 0$ and $T(v) = 1$. The objective aperture function $B(v)$ describes the effect of the objective aperture, transmitting or stopping electrons scattered in a particular direction v; $B(v)$ is unity for the open parts of the objective aperture and zero for the opaque parts. $B(v)$ therefore represents the fact that all electrons scattered at angles larger than $\alpha_{obj} = \theta_{max}$ (the semi-angle subtended by the objective aperture at the specimen) are stopped by the aperture and do not contribute to the image.

The maximum spatial frequency $v_{max} = \alpha_{obj}/\lambda$ determines the best resolution that can be achieved in the image; for example, if $\alpha_{obj} = 0.01$ rad (a 40 μm objective aperture for an objective lens of focal length 2 mm), $v_{max} = 2.7$ nm^{-1} for 100 keV electrons and the resolution in the image cannot be better than $1/v_{max}$, that is, 0.37 nm.

The image wavefunction $\psi_i(r)$ is calculated from the inverse Fourier transform of $S_i(v) = S(v)T(v)$; in the weak phase approximation:

$$S_i(v) = \delta(v) \exp\left[-i\gamma(v)\right]B(v) - iA(v)\cos\left[\gamma(v)\right]B(v) - A(v)\sin\left[\gamma(v)\right]B(v)$$
$$(3.20)$$

since $\exp(-i\gamma) = \cos\gamma - i\sin\gamma$.

The image wavefunction $\psi_i(r)$ calculated from Eq. (3.20) is in bright-field microscopy ($B(v) = 1$ for $v = 0$):

$$\psi_i(r) = 1 - i\eta(r)*q'(r) - \eta(r)*q(r) \qquad (3.21)$$

where q' and q are, respectively, the inverse Fourier transforms of $B\cos\gamma$ and $B\sin\gamma$. The asterisk $*$ represents the *convolution* integral:

$$f(r)*g(r) = \int f(r')g(r - r')\,dr' \qquad (3.22)$$

since a multiplication in Fourier space $F(v)G(v)$ becomes a convolution in real space $f(r)*g(r)$. The convolutions in Eq. (3.21) describe the blurring effect of the microscope lens aberrations on the specimen structure $\eta(r)$. The unity in Eq. (3.21) describes the effect of a uniform background in bright-field microscopy due to the unscattered electrons.

The image intensity $j(r)$, which is measured, is calculated from Eq. (3.21) by evaluating ψ_i times its complex conjugate ψ_i^+ (complex number $c = a + ib$; complex conjugate, $c^+ = a - ib$; $cc^+ = a^2 + b^2$):

$$j(r) = \psi_i(r)\psi_i^+(r) = |\psi_i(r)|^2 = 1 - 2\eta(r)*q(r) \qquad (3.23)$$

where the 'squared' or non-linear terms: $[\eta(r)*q'(r)]^2 + [\eta(r)*q(r)]^2$ have been neglected, consistent with the original approximation of neglecting terms including η^2. Equation (3.23) shows that in the weak phase approximation the image j is *linearly* related to the specimen structure and the image represents a distorted version of the structure as a result of the effect of the lens aberration function $q(r)$. This is the first important relationship that shows that for a thin specimen the image *does* represent the specimen provided that the lens aberration function does not seriously affect it. More informative in this respect, and forming the basis for the whole of image analysis in this book, is the *image transform*, which is obtained by taking the Fourier transform of the image intensity j in Eq. (3.23):

$$J(v) = \delta(v) - 2A(v)\sin[\gamma(v)]B(v) \qquad (3.24)$$

or, omitting the delta function: *The transform of an image of a thin specimen is the product of the transform of the specimen structure (η) and a microscope transfer function.* In the image transform there are two independent contributions, one dependent *only* on the specimen and the second dependent *only* on the electron microscope optics. The transfer function $\sin\gamma(v)$ depends on C_s, Δf and C_a; for the present it will be assumed that axial astigmatism can be corrected so that C_a is effectively zero. Then, depending on the value of C_s and the Δf chosen, $\sin\gamma$ will alter and the image transform will reflect these changes. In the ideal image $\sin\gamma$ should be made as near unity as possible, so that the image transform truly represents A, and hence the image seen on the micrograph represents true structural information. The effects of the microscope transfer function and its measurement are considered in detail in § 3.2.2.

Although a simple relationship between the image and the specimen can be derived in bright-field microscopy, the relationship in dark-field microscopy (Agar et al. 1974) is more complicated, suffering from the additional disadvantage that the image transform cannot be simply related to $A(v)$ or the microscope transfer function. Dark-field imaging is usually achieved by stopping the unscattered beam (the $\delta(v)$ in Eq. (3.20)) using a centre stop in the objective aperture (see § 5.4). Thus in Eq. (3.20) $B(v) = 0$ for $v = 0$, and the equation for the image intensity in dark-field microscopy comprises just the 'squared' terms which were neglected in the bright-field case, Eq. (3.23):

$$j(r) = [\eta(r)*q'(r)]^2 + [\eta(r)*q(r)]^2 \qquad (3.25)$$

The image transform will not give any useful information on the microscope transfer function. Also it can be seen from Eq. (3.25) that the image is not linearly related to η and this can cause problems in image interpretation (§ 5.4). Transforms of dark-field images must be treated with caution. Dark-field microscopy is considered in more detail later in this book: there is no uniform background to lower image contrast as in bright-field microscopy so there is potential for imaging unstained biological specimens with higher contrast.

3.2.1   Specimen thickness limitation in the weak phase approximation

The validity of the weak phase approximation depends on the neglect of terms in η^2 in the expansion of $\exp(-i\eta) = 1 - i\eta - \eta^2/2 + \ldots$ and the neglect of the 'squared' terms in the equation for the image intensity j.

In order to estimate the thickness limit for the weak phase approximation a knowledge of V is required for stained and unstained specimens. In order to obtain an indication of the magnitude of η it will be assumed that $V(r, z)$ does not change in the z-direction; thus from Eq. (3.13):

$$\eta(r) = \sigma \int_0^t V(r, z)\,dz \simeq \sigma t V(r) = 2.09 \times 10^{-3} \lambda t V(r) \qquad (3.26)$$

For unstained biological specimens the mean value of $V(r)$, V_m, is about 4 V, whereas for negatively-stained specimens V_m is about 16 V (Grinton and Cowley 1971); these figures are approximate but they will give an indication of the specimen thickness limits for which $\eta^2/2$ can be neglected in comparison with η. For example with 100 keV electrons ($\lambda = 3.7$ pm), $\eta = 7.73 \times 10^{-3} V_m t$ (V_m in volts, t in nanometres); for an unstained specimen $V_m = 4$ V, $\eta = 0.62$ rad, $\eta^2/2 = 0.19$ rad, for $t = 20$ nm, so that the error in the approximation is about 31 %. For a stained specimen ($V_m = 16$ V), the same error results for a specimen only 5 nm thick. For thicker specimens the error becomes so large that the weak phase approximation may not give a reliable representation of the image at all. The use of lower energy electrons increases λ and so further decreases the thickness limits for which the approximation can be used. However, in case these thickness limits look too restrictive, it should be pointed out that it is the variations of η across the biological structure that are important, rather than the absolute value of η. Thus if the mean value of the phase shift is η_m, then structurally it is only the variations about η_m, $(\eta - \eta_m)$ that are important. The weak phase approximation may still be valid provided that $(\eta - \eta_m)$ is significantly greater than $(\eta - \eta_m)^2/2$, and this is less restrictive in terms of specimen thickness, particularly for unstained biological specimens (§ 3.4.3).

The second approximation, the neglect of the 'squared' terms, can be evaluated from Fig. 3.2, where image contrast $\Delta j/j$ is plotted against the

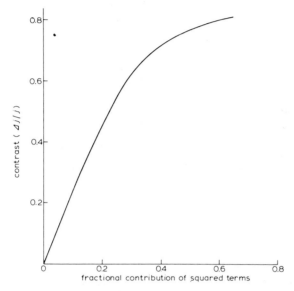

Fig. 3.2. The average contribution from the 'squared' terms to the image intensity in bright-field microscopy for an image contrast $\Delta j/j$. $E_0 = 100$ keV, $v_{max} = 2.7$ nm^{-1}, $\alpha_{obj} = 0.01$ rad.

fractional contribution (as compared with $2\eta^*q$) of the 'squared' terms to the image intensity. Image contrast is defined here as the variation in the image intensity divided by the background intensity, approximately $2\eta^*q/1$ in bright-field microscopy. For the normal contrast level of about 20% achieved in the microscopy of biological specimens the error due to non-linear terms is less than 10%. Figure 3.2 can be used to estimate the magnitude of the error that is likely to occur for a given image contrast. This is a relatively easy measurement to make on all micrographs, by using a microdensitometer to produce a few line scans across an area of the micrograph, and making a calculation of the magnitude of the variations in optical density divided by the mean optical density of the micrograph. A better way to estimate the validity of the weak phase approximation involves the numerical analysis of image transforms from a sequence of micrographs taken at incremental focus steps, *a focus series*, but this involves much more effort in addition to access to facilities for calculating numerical transforms (§ 3.4).

3.2.2   *The transfer function of the electron microscope*

The behaviour of the transfer function $\sin\gamma(v)$ of the electron microscope affects image interpretation and an incorrect choice of defocus Δf can lead to contrast reversals. The ideal transfer function would correspond to $\sin\gamma = 1$, but this can only be achieved over a limited spatial frequency (resolution) range.

Figure 3.3 shows the behaviour of $\sin\gamma$ near to focus. It is evident that the function oscillates between positive and negative values; negative values of $\sin\gamma$ mean that certain parts of the specimen transform will be multiplied by a negative number, completely reversing the contrast over the range of spatial frequencies for which $\sin\gamma$ is negative. Elsewhere, even if $\sin\gamma$ is positive, it will be less than unity and corresponding Fourier coefficients, A, will be attenuated. Thus the electron microscope can be considered to modify the *Fourier spectrum* of the specimen, and it is this distorted spectrum that will be evident in the image. The effect on the image will not be as easy to see as in the transform, but contrast reversals produced by altering defocus are clear indications of false detail in the image.

Images are never recorded overfocus, because the spherical aberration and defocus terms are additive and cause rapid oscillations in $\sin\gamma$. Images taken just underfocus (0–50 nm) have a small $\sin\gamma$ in the 0–1 nm^{-1} resolution region, so structural information in this resolution range will be attenuated (Fig. 3.3c and d). As underfocus is increased to 75–100 nm, $\sin\gamma$ is unity or near unity over a significant spatial frequency range. The defocus where $\sin\gamma \simeq 1$ over as large a spatial frequency range as possible is referred to as optimum defocus Δf_{opt}. Although the microscope transfer function is optimised for $\Delta f = 75$–100 nm, it will be seen that low frequency (low resolution) information up to 2 nm^{-1} (0.5 nm) is transferred to the image with a relatively small amplitude. In order to transfer this low resolution information by phase contrast it may be necessary to use larger underfocus values. As shown in Fig. 3.4, large underfocus values give larger values of $\sin\gamma$ at low spatial frequencies, but the rapid sign reversals of $\sin\gamma$ at higher spatial frequencies are unavoidable. In principle these sign reversals can be corrected by image processing. Probably the most important effect of the microscope transfer function is the occurrence of regions where $\sin\gamma$ is zero

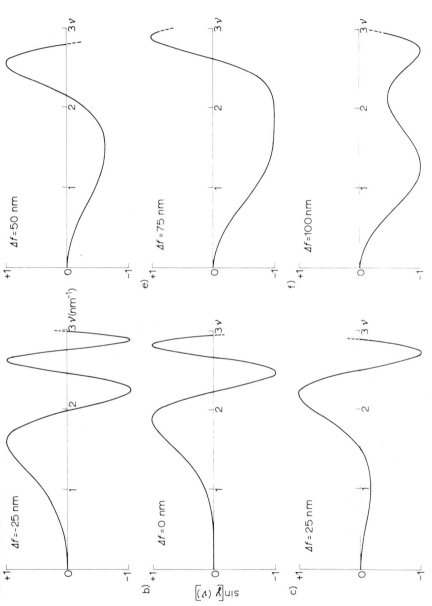

Fig. 3.3. The phase contrast transfer function $\sin\gamma(v)$ from (a) just over focus through (b) 'in focus' (c), (d) minimum contrast, approximately 25 nm underfocus to (e), (f) optimum defocus. $C_s = 1.6$ mm, $E_0 = 100$ keV. Coherence is assumed.

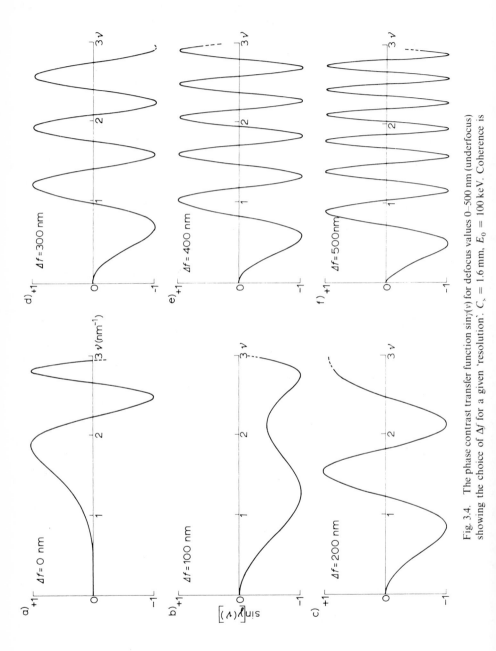

Fig. 3.4. The phase contrast transfer function $\sin\gamma(v)$ for defocus values 0–500 nm (underfocus) showing the choice of Δf for a given 'resolution'. $C_s = 1.6$ mm, $E_0 = 100$ keV. Coherence is

or near-zero; at these spatial frequencies no information on $A(\mathbf{v})$ is transferred to the image. The only way to restore this information is to take at least a second image of the specimen at a different defocus, and then to combine the two image transforms to restore the information missing in the frequency gaps (see § 4.4 and § 4.6).

If images are to be interpreted correctly, then it is essential that the correct defocus, and consequently the most suitable transfer function, is chosen. For example, if negatively-stained specimens are being used, it is unlikely that the biological structure is preserved to a resolution of better than 2 nm. Over the spatial frequency range 0–0.5 nm^{-1} the best defocus for optimising the phase contrast is about 500 nm underfocus (Fig. 3.4f). The rapid sign reversals of sin γ beyond about 1 nm^{-1} mean that detail in the image smaller than 1 nm cannot be believed. In the case of negatively-stained specimens the fine structure in the image mainly arises from the carbon support film; although evaporated carbon films are amorphous ($A(\mathbf{v})$ unstructured) they will appear granular as a result of the behaviour of the transfer function. For low resolution specimens it is necessary to use relatively large underfocus values if the detail in the specimen is to be imaged with maximum phase contrast. Sectioned material exhibiting a resolution of about 5–10 nm is not normally imaged by phase contrast because the large underfocus values required, approximately 3–14 μm, will lead to many spurious features in the image; such specimens are usually imaged near focus when the phase contrast is very small and the image contrast observed results from scattering contrast (see § 5.1). The approximate value of the defocus required to optimise phase contrast for a particular resolution can be calculated from Eq. (3.19) for sin γ; provided v is small (less than 1 nm^{-1}) the defocus term in Eq. (3.19) is significantly greater than the spherical aberration term. Thus for a spatial frequency v the transfer function can be optimised by choosing the Δf value that makes sin $\gamma = 1$ or $\gamma = \pi/2$, that is:

$$\sin\left[\frac{2\pi}{\lambda}\left(\frac{\Delta f v^2 \lambda^2}{2}\right)\right] = 1 \tag{3.27}$$

or for a spacing $r = 1/v$ in the specimen:

$$\Delta f = r^2/2\lambda \tag{3.28}$$

Hence for maximum phase contrast for the 2 nm spacing, with $\lambda = 3.7$ pm, $\Delta f = 540$ nm. Table 3.1 lists the values of Δf for a set of spacings between

TABLE 3.1

Approximate underfocus values Δf (μm) giving maximum phase contrast for a spacing r (nm) in the specimen, with electrons of wavelength λ (pm) and energy E_0 (keV)

r (nm)	E_0 (keV) λ (pm)	40 6.02	60 4.87	80 4.18	100 3.70
1		0.08	0.10	0.12	0.14
2		0.33	0.41	0.48	0.54
3		0.75	0.92	1.08	1.22
4		1.33	1.64	1.91	2.16
5		2.08	2.57	2.99	3.38
6		2.99	3.70	4.31	4.86
7		4.07	5.03	5.86	6.62
8		5.32	6.57	7.66	8.65
9		6.73	8.32	9.69	10.95
10		8.31	10.27	11.96	13.51
Δf_{opt} (μm) ($C_s = 1.0$–2.0 mm)		0.09 to 0.13	0.08 to 0.12	0.07 to 0.11	0.06 to 0.10
Δf_{min} (μm) ($C_s = 1.0$–2.0 mm)		0.04 to 0.08	0.03 to 0.06	0.03 to 0.05	0.03 to 0.05

Optimum defocus Δf_{opt} is defined in terms of the phase contrast transfer function for which $\sin \gamma \simeq 1$ over the largest spatial frequency range, and Δf_{min} corresponds to the defocus for minimum phase contrast.

1 nm and 10 nm; and a range of incident electron energies. The table also shows the values of Δf_{opt} for a range of C_s values. Δf_{opt} gives the best overall phase contrast image, when image detail up to a resolution of 0.4 nm ($v = 2.5$ nm^{-1}) can be believed, but detail of low resolution will be present with reduced contrast because of the small values of $\sin \gamma$ at low spatial frequencies. Choosing Δf to optimise a particular range of spacings, say 0–2 nm, will give good phase contrast for the chosen resolution range, but image detail smaller than the resolution corresponding to the first sign reversal in $\sin \gamma$ cannot be believed. If a crystalline specimen is being studied, then all the diffraction orders from the image must occur within the first zero of the transfer function; if certain diffraction orders occur in other than the first 'band' of $\sin \gamma$, this information will contribute to the image with the incorrect contrast; such an image cannot be interpreted reliably. Even if image processing is not of interest, the behaviour of the transfer function is still of interest so that images with the minimum amount of misinformation can be chosen for structural analysis.

The behaviour of $\sin \gamma$ near focus is also used as an aid for accurate focusing and correction of axial astigmatism. The use of Fresnel fringes around a hole in a carbon film for these purposes is not recommended if focus is to be determined with an accuracy much better than 1 μm. A carbon film owes its granular appearance in the image to the behaviour of sin γ. The minimum contrast observed in the carbon 'structure' occurs when sin γ is zero or near zero over a large spatial frequency range and this corresponds to the transfer function shown in Fig. 3.3c for $\Delta f = 25$ nm. The values of Δf_{min} for minimum contrast are also listed in Table 3.1; with the in-focus position determined, the fine or vernier focus control on the microscope can be turned the appropriate number of steps to achieve the required amount of underfocus. Also axial astigmatism can be corrected from the appearance of the carbon granular structure near focus, where the transfer function is most sensitive to changes in astigmatism. The direction of the streaky appearance of the carbon film changes orientation by 90° as focus is altered from just overfocus to underfocus. Astigmatism can be corrected so that there is no apparent rotation of the granular structure on going through focus (Agar et al. 1974).

3.3   Optical analysis of images

The simplest way to determine image transforms is to use an optical bench with laser illumination. Provided that optical image reconstruction facilities are not required, the quality of the optics and the optical bench is not important with the exception that a laser which produces a well collimated beam must be used. The micrograph can be considered to act as a complicated diffraction grating and the diffraction pattern obtained will reflect the information contained in the micrograph. It is important to remember that it is the image that is being examined and it is only by virtue of a relationship between the image and the specimen structure that structural information will be obtained from the image transform. For example, the image of a thick crystalline specimen may give an excellent diffraction pattern from which reliable measurements cannot be made, because of multiple electron scattering effects in forming the image. An optical transform can show whether an image can be reliably interpreted; for example, a biological specimen may consist of two layers of material, giving rise to a moiré pattern in the image, and provided that these two layers are rotated slightly with respect to each other, it will be immediately evident from the optical transform which will show closely spaced pairs of diffraction spots (§ 3.7.2; Fig. 3.39).

A simple optical diffractometer is shown diagrammatically in Fig. 3.5

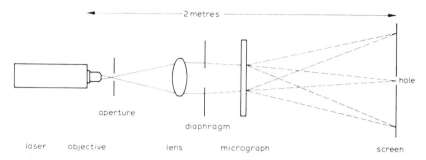

Fig. 3.5.   Diagram of an optical diffractometer for the analysis of electron micrographs.

(Horne and Markham 1972; Mulvey 1973). A He–Ne laser with an output of 2–5 mW is mounted at one end of a 2 m optical bench. A beam expanding system produces a beam of about 10–20 mm diameter. This means that diffraction patterns can be obtained from specimen areas as small as 100–200 nm in extent for images taken at an electron-optical magnification of 100,000 ×. For the beam expander, a 16 mm optical microscope objective is satisfactory used together with a 'spatial filter' which produces a clean beam profile; the spatial filter can be made from a 50 μm copper objective aperture mounted in a microscope objective holder. A convex lens of focal length about 0.25 m and of 0.1 m diameter is not essential but such a lens does enable the diffraction pattern from the micrograph to be focused and also enables the '*camera length*', and therefore, the size of the diffraction pattern, to be varied according to the type of specimen. For example, with low resolution spots or high magnification images the camera length can be increased to enlarge the diffraction pattern. A diaphragm can be used to select the area of the micrograph for analysis but it is not essential and usually the area can be altered by moving the micrograph carefully in its holder. The optical diffraction pattern is observed on a white card, often marked with circles as an aid to setting up the diffractometer for calibrating spacings; a small hole in the centre of the card allows the zero order beam to pass through the card and facilitates viewing the diffraction pattern, especially if the spots are weak or near the centre. A frosted glass screen ruled with a scale can be used for direct measurements on optical transforms. Normally optical diffraction patterns are photographed using either a 35 mm camera body, which has the advantage of giving a large number of exposures, or a Polaroid camera with a larger format. Exposure times vary, depending on the optical density of the micrograph and the size of the diffraction pattern; 0.1–2 sec is typical for Ilford FP4 film. Because of the relatively poor dynamic range of film, high intensity spots will saturate the photographic emulsion while low intensity spots may be lost in the fog. One of the main problems is the high intensity of the laser zero-order beam, but this can be attenuated by using a small black spot (1–2 mm) of Letraset or ink on a thin sheet of glass placed just in front of the camera of the diffractometer*. It is essential on all occasions to photograph the diffraction pattern of a calibration specimen, such as a copper grating, so that image transform spacings can be measured reliably. Optical diffraction patterns can be measured directly from the film using an eyepiece with a ruled scale, or the film can be photographically enlarged or even projected onto graph paper for measurement. With graph paper diffraction spots from a crystalline specimen are marked directly onto the paper; the calibration diffractogram is recorded at the same time, so the photographic enlargement need not be precisely known. The main problem with printing optical diffraction patterns is retaining the correct relative intensities of diffraction spots, because of the relatively poor dynamic range of photographic paper or film. Because only intensities can be recorded in an optical diffraction pattern, the phase information is lost. The problems of quantitative measurement of diffraction intensities and phase determination may be avoided by using numerical transforms (§ 3.4), but the optical diffractometer is ideally suited for the routine examination of electron micrographs. Vibrations of the optical bench are relatively un-important for recording diffractograms; this is because the amplitude of the Fourier transform is *translationally invariant*; that is, translational move-ments such as vibrations of the micrograph do not affect the positions or intensities of the spots (§ 3.4). Such a set-up will be unsuitable for optical reconstruction, because the reconstructed image is not translationally in-variant.

Micrographs to be examined on the optical bench must be dry, otherwise they will display diffraction patterns from the water still on the micrograph; they can be safely dried after dipping in alcohol if needed in a hurry. With the micrograph in place, the camera length can be adjusted by moving the lens and screen to obtain a diffraction pattern of suitable size for observation. Because of the inverse relation between real space and Fourier space, small spacings will give large diffraction spacings; thus a low magnification micrograph will require a smaller camera length than a high magnification micrograph of a particular specimen. Pictures with poor transforms, either by virtue of a poor transfer function (e.g. astigmatic) or blurred spots from

---

* *Warning*: it is dangerous to view the zero-order laser beam without suitable eye protection.

a crystalline sample, should be discarded. Transfer gaps can occur with a crystalline sample, leading to missing spots in the transform but provided the rest of the diffractogram is of good quality (sharp diffraction spots to high order), these micrographs should be retained, because a second micrograph may have this missing information; the two micrographs can be used together to form a composite transform for numerical image processing (§ 4.4).

3.3.1   The optical transform and optical diffractometry

The micrograph recorded at a magnification M acts as a diffraction grating for light of wavelength 632.8 nm (He–Ne laser). Thus an optical diffraction pattern can be used to determine the characteristic spacings in the micrograph and by implication the spacings in the original specimen. Mathematically the optical diffraction pattern corresponds to the Fourier transform of the micrograph. The interpretation of the optical diffraction pattern depends on the validity of the weak phase approximation and the linear relation between the image transform $J(v)$ and the transform $A(v)$ of the specimen structure $\eta(r)$, Eq. (3.24):

$$J(v) = \delta(v) - 2A(v)\sin\left[\gamma(v)\right]B(v)$$

However, it is only possible to record the intensity of the diffraction pattern, $|J|^2 = 4|A|^2\sin^2\gamma$, so that information on the phase of A is lost. Also the sign reversals of $\sin\gamma$ will not be observed; this ambiguity is not important because as soon as $\sin^2\gamma$ is seen to go through a minimum in intensity (zero of $\sin\gamma$), it is known that the next maximum in $\sin^2\gamma$ corresponds to a sign reversal of $\sin\gamma$. This loss of phase information means that the Fourier transform of an optical diffraction pattern cannot be used to determine $\eta(r)$. The inverse Fourier transform of $|A|^2$ is merely the *auto-correlation function* of η, that is:

$$\int |A(v)|^2 \exp\left(2\pi i v \cdot r\right) dv = \int \eta(r')\eta(r' + r)\, dr' \qquad (3.29)$$

The only thing that can be determined from $|A|^2$ is the symmetry of the crystal lattice and possible arrangement of the molecules in the structure; little can be said about molecular shape using $|A|^2$ alone.

Care should be taken in analysing the optical diffraction patterns of images of thick specimens or dark-field images. Both will contain contributions to the image from η^2, so that the optical transform J will contain terms such as:

$$\int \eta^2(\boldsymbol{r}) \exp(-2\pi i \boldsymbol{v} \cdot \boldsymbol{r}) \, \mathrm{d}\boldsymbol{r} = \int A^+(\boldsymbol{v}') A(\boldsymbol{v}' + \boldsymbol{v}) \, \mathrm{d}\boldsymbol{v}' \qquad (3.30)$$

the auto-correlation function of A. Then the optical diffraction pattern will not give reliable information on A or even $|A|^2$; this misinformation may be evident as additional diffraction spots that cannot be fitted into a simple

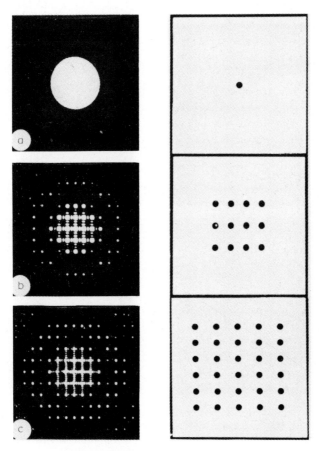

Fig. 3.6. Optical transforms: building up a lattice in two dimensions; subsidiary maxima are due to the limitation of the lattice. (From Harburn, Taylor and Welberry, Atlas of Optical Transforms, Bell and Hyman Ltd. 1975.)

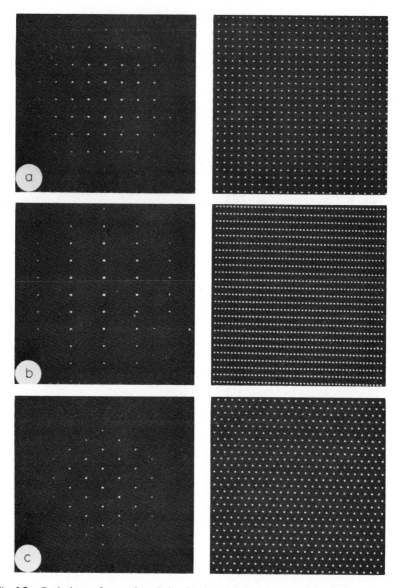

Fig. 3.7.   Optical transforms: the relationship between real space and the reciprocal lattice: (a) a square lattice, (b) a rectangular lattice, (c) a hexagonal lattice. Note the inverse relation between real space and reciprocal space. (From Harburn, Taylor and Welberry, Atlas of Optical Transforms, Bell and Hyman Ltd. 1975.)

lattice. Dark-field micrographs may give rise to spacings in the diffractogram which are one half of the actual lattice spacing as a result of the non-linear relation between j and η.

In order to give some indication of the use of the optical transform in the analysis of crystalline lattices, Fig. 3.6 shows an optical simulation of the build up of a lattice from a single motif (right-hand half of Fig. 3.6). As the lattice increases in size the diffraction spots become sharper (left-hand half of Fig. 3.6); thus as the number of units in a lattice increases, the diffraction pattern resolves into single discrete spots. Measurements can still be made on a diffraction pattern such as Fig. 3.6b, but it will not give a very accurate result for the lattice spacing. In this example a measurement could be made on the micrograph, but this would not usually be possible with a noisy electron micrograph of relatively low contrast. Figure 3.7 illustrates the relationship between the symmetry of the lattice (right-hand half) and the symmetry of the optical transform (left-hand half). In fact the lattice and reciprocal lattice are geometrically similar and the main difference is the reciprocal relation between the spacings. Thus the smaller horizontal spacing of the lattice in Fig. 3.7b becomes the larger horizontal spacing in the optical transform. The optical transform of a crystalline lattice shows the symmetry of the original lattice; often this will not be evident from the micrograph, but because of the discrete nature of the optical transform, even a noisy micrograph of a crystalline lattice should give visible diffraction spots.

3.3.2 Assessment of images: image quality

This section will consider how to evaluate the electron-optical quality of an electron micrograph from an examination of the transfer function of the image. This will not inform on the state of preservation of the specimen if it is non-crystalline, for its transform will be continuous and indistinguishable from the transform of the carbon support film. The main objective is to ensure that the image is not 'microscope-limited'; that is, that the correct choice of $\sin \gamma$ and Δf has been made and that the micrograph does not show astigmatism, specimen drift or any other image defects that will prevent a reliable visual interpretation. The quality of an image is usually assessed by taking the optical transform from a region of the carbon support film near the specimen or even from the specimen area, provided it is not too thick. The assumption is that the carbon support film satisfies the weak phase approximation so that its transform is proportional to $A \sin \gamma$. However, for evaporated carbon films A is relatively smooth so that the transform of such a

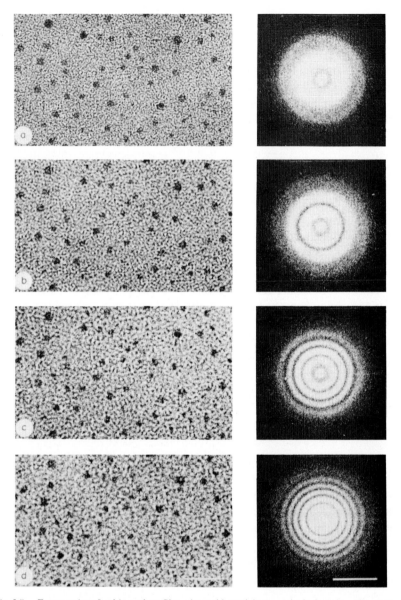

Fig. 3.8.   Focus series of a thin carbon film, plus gold particles to assist in focusing. The optical diffractograms are alongside each image: (a) optimum defocus, (b) 150 nm underfocus, (c) 210 nm underfocus, (d) 250 nm underfocus. C_s and Δf are determined by measuring the radii (or diameters) of the dark circles corresponding to minima in the optical transform. $E_0 = 125$ keV, image bar $= 10$ nm, diffraction bar $= 3.0$ nm^{-1}. (P. Sieber, unpublished.)

film gives essentially the shape of $\sin \gamma$ with a maximum spatial frequency $v_{max} = \alpha_{obj}/\lambda$. The optical transform intensity $|J|^2 = |A|^2 \sin^2 \gamma$ can then be used to assess microscope performance, the defocus of images and axial astigmatism, because these defects will show up in $\sin^2 \gamma$. Figure 3.8 shows a focus series of images of a carbon film starting at optimum defocus (Fig. 3.8a) in 50 nm defocus steps to about 250 nm underfocus (Fig. 3.8d), where the rapid oscillations of $\sin \gamma$ are evident in the corresponding optical transform. Only $\sin^2 \gamma$ is determined, but because it is known that $\sin \gamma$ alternates in sign, alternate bright rings (white on print, black on negative) correspond to changes in the sign of $\sin \gamma$. Dark rings correspond to minima in $\sin \gamma$ where little or no information is present on the specimen structure at these spatial frequencies. If the frequency scale of the diffraction pattern is determined in terms of v (nm^{-1}) the positions of these minima can be equated with the zeros of $\sin \gamma$, thus enabling Δf and C_s to be determined. Astigmatism can also be determined from the distortion of the rings into ellipses (see below). The series shown in Fig. 3.8 is excellent in this respect, because the diffraction patterns are almost perfectly circular. Even if measurements of Δf and C_s are of no interest, an examination of a focus series of micrographs on an optical diffractometer provides an efficient means of choosing those micrographs which were taken at the correct setting of defocus.

It will be seen that the optical transforms in Fig. 3.8 do not cut off sharply at the objective aperture $(v_{max} = 3 \text{ nm}^{-1})$ but the intensity gradually falls off towards the outer regions of the diffractogram. This attenuation of the higher spatial frequencies is due to the limited coherence of the electron source; this does not usually affect image transforms of biological specimens which seldom extend beyond 1 nm^{-1}. However, the effects of limited spatial coherence (angular spread of the incident beam, $\beta = v_0 \lambda$, in Fig. 3.1; § 3.2) and limited chromatic coherence (energy spread of the incident beam) should always be minimised by the correct operation of the electron microscope. Spatial coherence can be improved by using a relatively small condenser lens aperture (50–100 μm) and working well away from the second condenser lens cross-over; the chromatic coherence may be increased by using a lower filament temperature. But these efforts to improve source coherence cannot be extended too far since they will lead to a significant loss in beam intensity and consequent problems in focusing and correcting astigmatism.

Figure 3.9 shows the profiles of $\sin^2 \gamma$ corresponding to the focus series in Fig. 3.8, and it is evident that the attenuation of the $\sin^2 \gamma$ curves causes loss of information beyond a spatial frequency of 2 nm^{-1} (0.5 nm resolution); the parameters used to derive the curves in Fig. 3.9 were an angle of illumina-

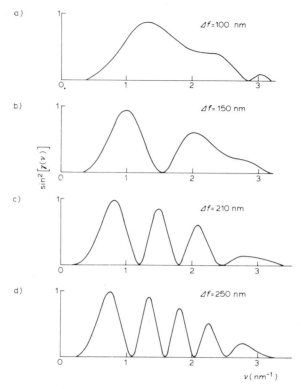

Fig. 3.9. Theoretical results for the intensities $\sin^2\gamma(v)$ of the optical transforms of the carbon film series shown in Fig. 3.8. Coherence envelope on $\sin^2\gamma$ corresponds to an illumination angle β of 0.5 mrad with a thermal energy spread E_f of 1 eV. $C_s = 1.9$ mm, $C_c = 1.6$ mm, $E_0 = 100$ keV.

tion of $\beta = 0.5$ mrad and a filament energy spread of $E_f = 1$ eV. Clearly such imaging conditions are not going to seriously affect images of biological specimens, but the microscopist should be aware that using a microscope with the illumination near cross-over and an overheated filament can lead to a serious loss of high resolution information (Beorchia and Bonhomme 1974). This effect is shown in Fig. 3.10 where E_f is increased from 0 to 6 eV; clearly for the largest value of E_f spatial frequencies as low as 1 nm^{-1} are attenuated by the limited chromatic coherence of the electron source.

It will be evident that, although it is desirable to use and know accurately the electron-optical conditions under which the image is recorded, this should not be emphasized at the expense of giving a large radiation dose to the specimen. The specimen should be photographed at minimal or even sub-

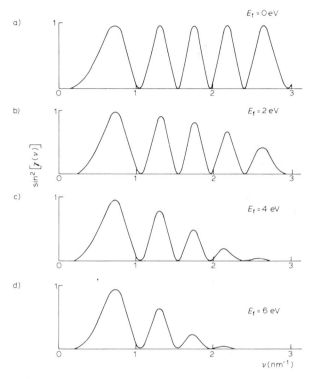

Fig. 3.10. Theoretical curves showing the effect of beam incoherence on the intensity $\sin^2\gamma(v)$ of the optical transform of a weak phase object at defocus $\Delta f = 250$ nm and $E_f = $ (a) 0 eV, (b) 2 eV, (c) 4 eV, (d) 6 eV. $C_S = 1.9$ mm, $C_C = 1.6$ mm, $E_0 = 100$ keV.

minimal doses, followed by a second image of longer exposure which can be used to determine $\sin \gamma$ (Unwin and Henderson 1975); subminimal dose images will usually have too low an optical density on the film or plate to give a clearly visible carbon optical transform. In addition to obtaining a measure of Δf of the image, optical diffractometry can be used routinely to detect other image defects. Axial astigmatism shows up as an ellipticity of the ring pattern, because its effect is equivalent to a defocus difference of approximately C_a in the direction of the astigmatism. When Δf is small, astigmatism shows up most clearly because the C_a term is dominant in Eq. (3.19) for $\sin \gamma$. For large values of defocus, around 500 nm, even an astigmatic error equivalent to a defocus difference of 100 nm, will barely distort the rings. Figure 3.11 shows several images of carbon films with varying degrees of astigmatism (note $\Delta z \equiv \Delta f$). The difference between the defocus values, $\Delta f_1 - \Delta f_2$,

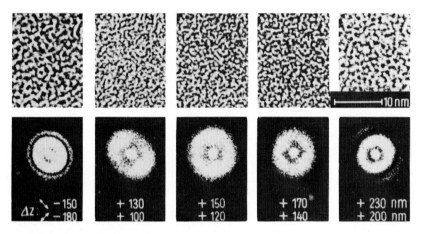

Fig. 3.11.   Micrographs and optical transforms of a carbon film, showing the effect of axial astigmatism and the determination of its direction. (P. Sieber, unpublished.)

measured along the major and minor axes of the ellipse gives a measure of the astigmatism. Whether or not an image should be rejected as a result of astigmatism depends on the relative magnitudes of Δf_1 and Δf_2; a micrograph taken when Δf is much greater than C_a is usually acceptable. In fact the images shown in Fig. 3.11 are not highly astigmatic, with $|\Delta f_1 - \Delta f_2| = 30$ nm. A large amount of axial astigmatism will cause the maxima in $\sin \gamma$ to be in completely different positions to the maxima at right angles and this will cause information from the specimen to be distorted. In extreme examples, usually visible in the micrograph, the astigmatism will be large enough to give an optical transform with the appearance of a Maltese cross or the micrograph may not even give a transform. In a high resolution electron microscope axial astigmatism can normally be corrected at high magnification (200,000 ×) on the carbon film, with an error of about ± 20 nm. The magnification can then be reduced to that used for taking pictures without altering the astigmatism. The use of thin foil apertures or even dispensing with an aperture in the objective lens, avoids problems of a continuously changing astigmatism caused by a contaminated objective aperture; this removal of the objective aperture is recommended only for phase contrast microscopy, because scattering contrast images depend on the objective aperture for their differential contrast (§ 5.1).

Other image defects will show up as various distortions of the ring pattern. In Fig. 3.12a, specimen drift shows as a cut-off of the transform at right-angles to the direction of drift; this defect is often unavoidable with minimum

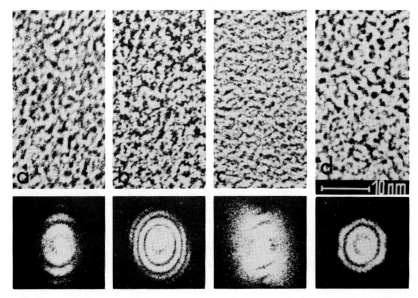

Fig. 3.12. Micrographs and optical transforms of a carbon film, showing various image defect: (a) specimen drift, (b) miscentred objective aperture, (c) and (d) electrical charging of the objective aperture as a result of, for example, contamination. (P. Sieber, unpublished.)

radiation techniques but it does mean that structural information in one direction is blurred and completely lost in the transform. Fig. 3.12b shows the effect observed when the objective aperture is misaligned so that the outer parts of the transform are cut off, and Fig. 3.12c and d show the effects of electrical charging of the aperture, for example, as a result of a contaminated aperture.

Although the image is of a three-dimensional structure, the information in the optical transform is essentially two-dimensional. However, under certain conditions three-dimensional information is evident: for example, a *focus gradient* resulting from variations in the surface topography of the carbon support film will show as a change in the ring pattern as the micrograph is moved across the laser beam. Micrographs with large focus gradients may have to be discarded, because the spatial information in the image will vary in the direction of the gradient. But if the appearance of the molecules in the image appears to be independent of this focus gradient the effect may be unimportant. Micrographs of tilted specimens will have focus gradients and corrections should then be applied for the changes in the transfer function (Henderson and Unwin 1975). A thick specimen also causes problems because the transfer function at the top of the specimen will be different from that at the bottom, equivalent to a focus difference $\Delta f = t$. This shows up in the transform as an apparent loss of high frequency information, particularly at large defocus values, where the zeros of one $\sin \gamma$ will be filled in by the non-zero $\sin \gamma$ from a different level in the specimen. Such images cannot be completely corrected for the effects of the transfer function. The appearance of the transform of such an image will assist in deciding whether to proceed with further analysis of the image either visually or by image processing. Efforts should be made to prepare thinner specimens which are amenable to image analysis.

3.3.3   Assessment of images: image resolution

The general characteristics of an image can be assessed by optical diffracto-metry without knowing the precise spatial frequency range of the optical diffraction pattern. But for the determination of image defocus from the optical transform of a carbon film or the spacing and resolution of a crystalline specimen, the diffractometer will need to be calibrated using a standard spacing. The determination of the lattice parameters of a biological specimen may be important in assessing changes in a system, such as the effect of removal or addition of lipid to membrane systems. Whilst in principle this could be determined by tedious measurements on the micrograph, it is much easier to measure, not only the lattice spacings, but also the order/disorder of the specimen from its optical diffraction pattern. A calibration is required for every set of optical diffraction patterns recorded. One method of calibration is to record an image of a negatively-stained catalase crystal (Wrigley 1968) at the same magnification M as the test images. This has the advantage of eliminating the need to know M exactly. The reciprocal spacing (usually measured in nm^{-1}) is given by:

$$v = Cd \qquad (3.31)$$

where C is the diffractometer constant and d is the distance (usually measured in mm) of a diffraction spot from the centre of the diffraction pattern. d is usually obtained as an average value from the measurement of several diffraction spots (see below). Since the catalase 'a' and 'b' spacings are known, C can be determined from measurements on the diffractogram of the catalase image: $v_{cat} = Cd_{cat}$. Figure 3.13 shows the images and respective optical diffraction patterns of (a) thin and (b) thick negatively-stained

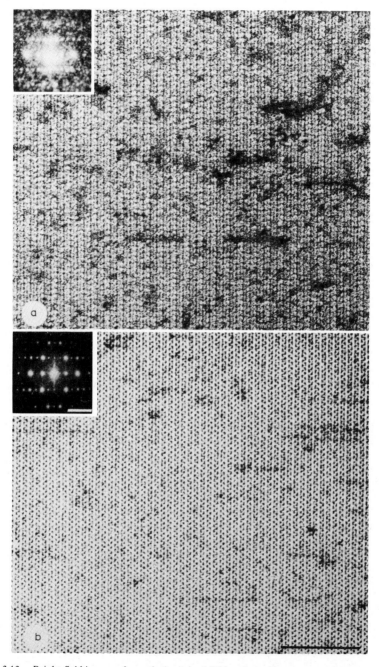

Fig. 3.13. Bright-field images of negatively-stained (2% sodium silicotungstate, SST), glutaraldehyde-fixed catalase crystals, together with their optical transforms: (a) thin crystal, (b) thick crystal. Image bar = 100 nm, diffraction bar = 0.2 nm^{-1}, E_0 = 80 keV. (N.G. Wrigley and E.B. Brown, unpublished.)

catalasé crystals. The horizontal 'a' spacing is 17.5 nm which corresponds to twice the most evident repeat distance in the micrographs in Fig. 3.13; thus $v_{cat} = 1/17.5 \, nm^{-1}$ for the horizontal separation of diffraction spots in the optical transforms shown as insets; the vertical 'b' spacing, which is 6.9 nm, is not so easy to measure on the micrograph but can be used as a calibration check on the optical diffractometer with $v_{cat} = 1/6.9 \, nm^{-1}$. Figure 3.14

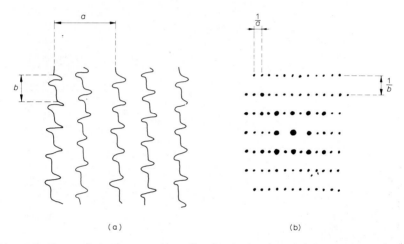

(a)   (b)

Fig. 3.14.   The use of a catalase crystal for calibration. (a) the schematic structure of a negatively-stained crystal, (b) its optical diffraction pattern.

shows the relationship between (a) the catalase lattice and (b) the optical diffraction pattern; this shows why the 'a' spacing in the micrograph appears to be 8.75 nm but is in fact twice this value because each row of catalase molecules is displaced vertically by one half a unit from the previous row, so that the true repeat distance spans two rows. Several measurements of $1/a$ can be made on the diffraction pattern and it is advisable to determine $1/a$ by measurement across at least 10 diffraction orders to calculate the average $1/a$ value. Catalase can also be used to calibrate the microscope magnification; accurate values for M are required when the image transform is determined numerically, since the normal magnification values given in the microscope handbook may be in error by as much as 10%.

The second method of calibration of optical diffraction patterns uses a standard copper grating or a fine mesh microscope grid whose spacing can be measured optically with a travelling microscope. However, in this procedure the image magnification must be determined. The fragile copper grating (spacing about 40 lines/mm) is mounted between thin glass plates for support. A comparison is then made in the optical transform of the grating spacing and the micrograph spacing ($M \times$ specimen spacing). Setting $v_{gr} = L \times 10^3$ lines/m in $v_{gr} = Cd_{gr}$ (Eq. (3.31)) where d_{gr}, the spacing between successive diffraction orders from the grating, is set to about 25 mm (to conveniently fit onto 35 mm film). Now for the micrograph, the spatial frequency $v_m = v/M$, where v is the actual reciprocal spacing in the specimen (note the inverse relationship to image spacing $= M \times$ specimen spacing); thus, using $v_m = Cd_m$, the actual reciprocal spacing in the specimen can be calculated from:

$$v = \left(\frac{L \times M}{d_{gr}} \times 10^3\right) d_m$$

The units of v are then m^{-1}; it is more usual to calculate v in units of nm^{-1}: if L is in lines/mm, M is in units of 1000, then:

$$v = \left(\frac{L \times M \times 10^{-3}}{d_{gr}}\right) d_m \tag{3.32}$$

The 'real' spacing in the specimen r can then be determined from $1/v$ nm. For example, using a grating with $L = 40$ lines/mm, the image of a neuraminidase crystal taken at $M = 40,000$ gave a reciprocal spacing of $d_m = 1.6$ mm measured from the optical diffractometer film; the grating calibration gave $d_{gr} = 25$ mm. The diffractometer constant in Eq. (3.32) is then $40 \times 40 \times 10^{-3}/25 = 0.064$, and $v = 0.064 \times 1.6 = 0.1024$ nm^{-1}, corresponding to a real 'a' spacing of 9.8 nm (see § 3.7; Figs. 3.30 and 3.31).

It should be noted that in order to determine the 'a' and 'b' spacings of a crystalline lattice, the symmetry of the lattice must be known and all or most of the diffraction orders must be fitted onto a single reciprocal lattice, with diffraction spots at (ha^*, kb^*) for a two-dimensional crystal, where h and k are integers and a^* and b^* are the reciprocal lattice constants (§ 3.1.2). In two dimensions, a reciprocal lattice is defined by a^*, b^* and the angle α between the principal axes (Beeston 1972). Thus for a square lattice the spatial frequency corresponding to diffraction order (h, k) is:

$$v_{hk} = a^*\sqrt{h^2 + k^2}, \qquad a^* = b^*, \alpha = 90° \text{ and } a = 1/a^* \tag{3.33}$$

Neuraminidase crystals have a square lattice with $a = b = 9.8$ nm (Fig. 3.30).

For a rectangular lattice in two dimensions:

$$v_{hk} = \sqrt{h^2 a^{*2} + k^2 b^{*2}}, \qquad a^* \neq b^*, \alpha = 90°, a = 1/a^*, b = 1/b^* \qquad (3.34)$$

For thin catalase platelets, $a = 17.5\,\text{nm}$, $b = 6.9\,\text{nm}$ (Figs. 3.13 and 3.14). For a hexagonal lattice:

$$v_{hk} = a^* \sqrt{h^2 + hk + k^2}, \qquad a^* = b^*, \alpha = 60°, a = \sqrt{4/3} \times 1/a^* \qquad (3.35)$$

Monolayers of hexons from adenovirus form a hexagonal lattice, with $a = 22\,\text{nm}$ (Fig. 3.15).

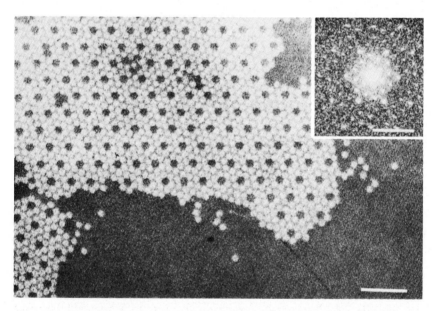

Fig. 3.15.   Electron micrograph and optical diffraction pattern of negatively-stained (2 % phosphotungstic acid, PTA) monolayers of hexons from adenovirus. Image bar = 50 nm, diffraction bar = $0.2\,\text{nm}^{-1}$, $E_0 = 60\,\text{keV}$. (From Pereira and Wrigley 1974.)

All, or at least the most intense, diffraction orders should fit one of these equations, provided that the *c*-axis of the crystal is perpendicular to the incident electron beam ($l = 0$). The process of choosing a reciprocal lattice to fit the diffraction spots is known as *indexing* the lattice, or *indexation*, and this is considered in more detail in § 3.7 because it is one of the most important steps in proceeding with image reconstruction; incorrect indexation can lead to a completely false reconstruction.

Information on the third dimension c of the crystal can be obtained by tilting the specimen, and then a complete description of the crystalline lattice and the arrangement of the molecules in the unit cell can be determined (see International Tables for X-ray Crystallography, Vol. I). However, much useful information can be obtained from the analysis of two-dimensional diffraction patterns. For example, the spatial extent of the diffraction spots indicates the preservation of the crystalline lattice and the degree of order/disorder of the subunits making up the crystalline lattice; these two effects can be summarised by quoting an image resolution corresponding to the reciprocal of the highest spatial frequency measured from the optical diffraction pattern. If the specimen is not ordered then the optical transform will not be discrete but continuous, showing no strongly recognisable features and then the specimen transform cannot be distinguished from that of the carbon support film. Nothing can then be said about specimen preservation; only the electron-optical quality of the image and image defocus can be determined (see § 3.3.4). The numerical transform is not quite so limited because specimens showing rotational symmetry can be assessed from the

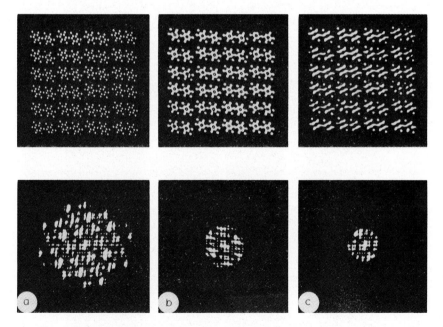

Fig. 3.16. Optical transforms: the relationship between resolution and the extent of the transform for a perfect crystal: (a) original transform, (b) and (c) spatially filtered results. (From Harburn, Taylor and Welberry, Atlas of Optical Transforms, Bell and Hyman Ltd. 1975.)

phase information $\omega(\mathbf{v})$ in the specimen transform $A(\mathbf{v})$. For example, for *n*-fold rotational symmetry the amplitudes and phases in the transform should be related by an *n*-fold rotational axis (see § 3.4, but mainly § 4.1).

Some optical simulations give an indication of the information that can be obtained from the optical diffraction patterns of ordered specimens (Figs. 3.16–3.18, from Harburn et al. 1975). The inverse relationship between real and Fourier space means that high spatial frequencies refer to fine detail in the image, whilst low spatial frequencies refer to coarse structure in the image. Thus in Fig. 3.16 the upper half shows a series of ordered structures with a progressive degradation in the fine detail of the motif; this corresponds exactly to the contraction of the optical diffraction pattern from Fig. 3.16a to Fig. 3.16c. Conversely an image reconstructed from a transform of small spatial extent will show only the coarse detail of the motif. Such an effect could be observed in electron microscopy as a result of the effect of the negative stain or radiation damage on the molecules in the crystal. Figure 3.17 shows that increasing the disorder of a lattice (right-hand half of figure) causes not only an attenuation of the diffraction pattern (left-hand half), but also a broadening of the spots, because the motifs no longer occur at exactly integral multiples of *a*. Measurements of such a diffraction pattern will give an average value of the lattice constant *a* whilst the width of the diffraction spots will give a measure of the variation (standard deviation) from *a*. The effect shown in Fig. 3.17 may be due to breaking the relatively weak bonds between molecules that form the crystalline array during staining and irradiation. In practice it may be difficult to distinguish between the effects shown in Figs. 3.16 and 3.17; that is, to differentiate between intramolecular and intermolecular damage of the biological specimen. However, such an analysis of images does enable a qualitative assessment to be made of the effect of different specimen preparation methods on the integrity of the structure – it should also be possible to distinguish between images taken under normal and minimal radiation conditions. Also Fig. 3.18 shows that it can reveal the presence of small crystals arranged so that they are (a) aligned (b) randomly arranged (polycrystalline) or (c) preferentially oriented. Because optical diffraction is a post-diagnostic tool, only usable after a picture has been taken, removed from the microscope, developed and dried, it should be used together with electron diffraction techniques (Beeston

Fig. 3.17. Optical transforms: the effect of lattice disorder on the diffraction pattern: (a) ordered, (b) $\pm 10\%$ vertical disorder, (c) $\pm 10\%$ two-dimensional disorder, (d) $\pm 25\%$ two-dimensional disorder. (From Harburn, Taylor and Welberry, Atlas of Optical Transforms, Bell and Hyman Ltd. 1975.)

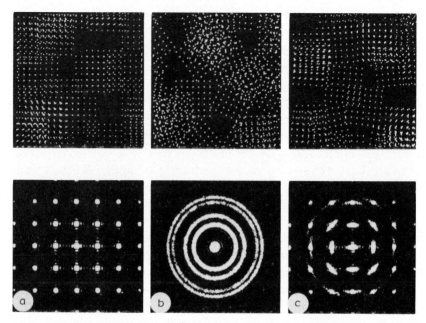

Fig. 3.18.   Optical transforms: a paracrystalline lattice: (a) random position of crystallites, but same orientation, (b) arbitrary orientation, (c) preferred orientation. (From Harburn, Taylor and Welberry, Atlas of Optical Transforms, Bell and Hyman Ltd. 1975.)

1972). In many cases the extent of the ordered structure may be too small to obtain a good electron diffraction pattern even by selected area techniques. Because the image contains an enlarged version of the structure, the area of specimen from which an optical diffraction pattern can be obtained may be as small as 100–200 nm in extent. For example, in the image of a neuraminidase crystal taken at 50,000 ×, an optical diffraction pattern may be obtained from an area $10 \times 10 \text{ mm}^2$ on the micrograph, equivalent to a specimen area of $0.2 \times 0.2 \mu\text{m}^2$; this is just 20×20 unit cells, too small to give a satisfactory electron diffraction pattern. So optical diffraction should be most useful in examining images of biological structures for the detection of ordered arrangements of the molecules, which may not be at all evident in the image. Besides the advantage of being able to examine smaller areas of the specimen by optical diffraction, an image is necessary for structural determination; the electron diffraction pattern contains information only on the magnitude $|A|$ of the specimen transform A, whilst the image transform contains information on both $|A|$ and the phase of A, ω, provided the specimen is thin. In both optical and computer reconstruction of the image this phase information is retained in the transform, although it cannot be recorded directly in an optical system.

Optical diffractometry may also be used to analyse specimens with one-dimensional order, such as collagen or muscle fibres. The optical transform will consist of streaks with a spacing characteristic of the one-dimensional repeat distance in the specimen. Thus, variations in the average spacing in the fibres can be determined by measuring the optical diffraction pattern.

3.3.4 Measurement of defocus from optical transforms

It may be sufficient to verify that the transfer function of the electron microscope does not distort the structural features in the image; that, for example, in a crystalline lattice all the diffraction spots fall within the first minimum of the transfer function. However, if the diffraction orders fall in different bands of the transfer function, there is the choice of discarding the micrograph or making a correction for the effect of the transfer function; in the latter case it will be necessary to determine the main parameters, C_s and Δf, of the transfer function $\sin \gamma$. If high resolution images are obtained then it will be virtually impossible to record images which are not adversely affected by the transfer function, because of the relatively fast variations in $\sin \gamma$ at high spatial frequencies. Axial astigmatism, if present, can be determined by measuring the defocus of the transfer function in two directions at right angles. The simplest way to determine C_s and Δf is to measure the positions of the minima in the optical diffraction pattern of the carbon support film (Fig. 3.8; § 3.3.2); these correspond approximately to the zeros of $\sin \gamma$ and the positions of these zeros will fit a particular range of C_s and Δf values in Eq. (3.19). Because the optical transform occurs on a steep background resulting from inelastic scattering, the positions of the minima are sometimes difficult to determine with precision. In principle, only two minima are required in Eq. (3.19) to determine C_s and Δf, but in practice at least 4 or 5 minima are required to determine the microscope parameters. This condition is impossible to satisfy for micrographs taken near optimum defocus (~ 100 nm underfocus) because the third zero is usually beyond the resolution of the microscope. Table 3.2 shows the first five zeros of $\sin \gamma$ for a defocus range of -100 to 500 nm ($C_s = 1.4$ mm, $\lambda = 3.7$ pm). If the value of C_s is taken from the manufacturer's data for the microscope, Δf is determined by matching the zeros measured from the optical transform (usually the diameter of the ring is measured to avoid centering errors) as nearly as possible with the corresponding zeros in the table. Often the measured zeros

TABLE 3.2

The first five zeros of the phase contrast transfer function $\sin[\gamma(v)]$ (excluding $v = 0$) for defocus Δf(nm), an objective lens with a spherical aberration coefficient C_s of 1.4 mm and an incident electron energy of 100 keV ($\lambda = 3.70$ pm)

Defocus Δf(nm)	Zeros of $\sin\gamma$, v (nm^{-1})				
	1	2	3	4	5
−100	1.49	1.98	2.31	2.57	2.78
− 75	1.64	2.13	2.46	2.72	2.93
− 50	1.82	2.31	2.64	2.88	3.09
− 25	2.04	2.51	2.82	3.06	3.26
0	2.30	2.74	3.03	3.26	3.44
25	1.61	2.60	2.99	3.25	3.46
50	2.28	2.92	3.25	3.49	3.68
75	2.80	3.24	3.52	3.73	3.90
100	3.23	3.56	3.79	3.97	4.13
125	1.65	3.21	3.61	3.86	4.06
150	1.44	2.37	3.16	3.68	3.95
175	1.30	1.98	3.78	4.07	4.27
200	1.20	1.79	2.35	3.92	4.20
225	1.13	1.65	2.11	2.60	4.09
250	1.06	1.54	1.95	2.34	2.76
275	1.01	1.46	1.83	2.17	2.51
300	0.96	1.38	1.73	2.04	2.33
325	0.92	1.32	1.65	1.93	2.20
350	0.89	1.27	1.58	1.84	2.09
375	0.86	1.22	1.51	1.77	2.00
400	0.83	1.18	1.46	1.70	1.92
425	0.80	1.14	1.41	1.65	1.86
450	0.78	1.11	1.37	1.59	1.79
475	0.76	1.08	1.33	1.55	1.74
500	0.74	1.05	1.29	1.50	1.69

v_0 will span a range of defocus. Ambiguities in Δf as a result of using too few values of v_0 can only be resolved using a focus series, with known incremental steps. It is unlikely that a focus series will not resolve the ambiguity between, say $\Delta f = -100$ nm and $\Delta f = 150$ nm using only two values for v_0, because the next micrograph transform in the series (say $\Delta f + 100$ nm) will distinguish between 0 nm ($-100 + 100$) and 250 nm ($150 + 100$) even with only two values for v_0. Unfortunately the focus increments given for most microscopes are not too reliable ($\pm 20\%$) and so the focus steps in a series may not be equidistant. Since the values of v_0 depend on both C_s and λ, tables such as Table 3.2 must be calculated for a range of C_s and incident electron energies; in practice a few such tables with a small range of C_s

(1–2 mm) and smaller increments in Δf than shown may be easily calculated using either a programmable calculator or a computer. The alternative of using a least squares curve fitting method to determine C_s and Δf is not recommended, particularly when only 2 or 3 values of v_0 are available.

Zeros of the transfer function are obtained whenever $\sin \gamma = 0$ or $\gamma = n\pi$; that is, from Eq. (3.19):

$$\frac{2\pi}{\lambda}\left(\frac{C_s v_0^4 \lambda^4}{4} - \frac{\Delta f v_0^2 \lambda^2}{2}\right) = n\pi \tag{3.36}$$

where n is an integer (positive, negative or zero). Equation (3.36) can be solved for v_0 in terms of C_s and Δf to give:

$$v_0^2 = \frac{\Delta f \lambda \pm \sqrt{\Delta f^2 \lambda^2 - 2nC_s\lambda^3}}{C_s\lambda^3}$$

and

$$v_0 = \left[\frac{\Delta f \lambda \pm \sqrt{\Delta f^2 \lambda^2 - 2nC_s\lambda^3}}{C_s\lambda^3}\right]^{1/2} \tag{3.37}$$

Now in order to generate a table such as that shown in Table 3.2, Δf and C_s are chosen and v_0 values are generated for a series of n values. However, there are restrictions on the values for n because all numbers under a square root sign must be positive (negative values will give imaginary values of v_0). Whether the positive or negative sign ($\pm$) is taken also depends on the size of $\sqrt{(\Delta f^2 \lambda^2 - 2nC_s\lambda^3)}$ in comparison with $\Delta f \lambda$; the difference must always be positive. If $\Delta f > 0$ (underfocus), Eq. (3.37) is solved starting with $n = +1$ and the decision whether to increase or decrease n by 1 depends on whether $(\Delta f^2 \lambda^2 - 2nC_s\lambda^3)$ is greater or less than zero. If $\Delta f < 0$ (overfocus), the equation is solved commencing with $n = -1$. Equation (3.37) is more useful if 'natural' units are used for C_s (mm), Δf (nm), λ (pm), when $2nC_s\lambda^3$ becomes $2,000nC_s\lambda^3$ and the rest of the equation is unaffected. The mathematics is given here because sets of tables, such as Table 3.2, are useful for obtaining an estimate of Δf for the particular microscope and accelerating voltage used.

If the image is astigmatic, two sets of v_0 will be obtained along the principal axes of the elliptical diffractogram and the two values of Δf, Δf_1 and Δf_2 will give the astigmatic difference $\Delta f_1 - \Delta f_2$. For example, an astigmatic difference of 50–100 nm is acceptable for an image taken 500 nm underfocus, but unacceptable for an image recorded 200 nm underfocus. Astigmatic correction is much more critical for high resolution images because $\sin\gamma$ varies much more rapidly at high spatial frequencies (γ depends on $C_a v^2$). Note that the lower order zeros of $\sin\gamma$ are relatively insensitive to variations in C_s, because the spherical aberration depends on $C_s v^4$ and C_s can only be reliably determined from high frequency information. Several tables with different C_s values and a suitable defocus range should be used together to find the value of C_s that gives the most consistent fit to experimental data.

In one sense the precise values of C_s and Δf do not matter because it is only the shape of the transfer function $\sin\gamma$ appropriate to an image that is required if an image is to be corrected for the effect of lens defects. There is no point in determining Δf, nor is it generally possible, to a precision that is better than the thickness of the support film plus specimen, from which the optical diffraction pattern was obtained. A thin carbon support film should be used because determining $\sin\gamma$ or Δf from the optical transform depends on the validity of the weak phase approximation giving an image transform proportional to $\sin\gamma$; thick carbon films give a transform intensity which includes contributions from $\cos\gamma$ which will displace the minima to higher spatial frequencies and in some cases make an accurate measurement of Δf impossible.

3.4   Computer analysis of images

In order to calculate a numerical Fourier transform, the density variations in the image must first be converted to *digital* information using a two-dimensional scanning microdensitometer. A numerical transform is not just used as an alternative to the optical transform; it provides additional information, such as the phases of the Fourier coefficients, and quantitative information on the amplitudes of the diffraction spots for a crystalline sample. It is also a prerequisite for certain types of image processing, such as rotational filtering (§ 4.1) and the straightening of a distorted lattice (§ 4.2.3). Producing a numerical transform requires not only a scanning densitometer, but also a computer with the necessary programmes and a suitable means of displaying the final results in a form that is visually acceptable.

3.4.1   Densitometry and digitising micrographs

There are two types of scanning densitometer, the flat-bed type and the rotating drum type. The former densitometer consists of a table on which the micrograph is mounted which moves in a two-dimensional raster with incremental steps as small as 1 μm. The density variations are determined by measuring the transmission through the micrograph of a small spot of light, using a photomultiplier. The rotating drum type densitometer has the micrograph mounted round a cylinder which rotates on a screw thread. As the drum rotates the light transmission through the micrograph is measured. The screw thread moves the micrograph in front of the fixed light spot to produce a two-dimensional scan. The micrograph must be on film, whereas both plates and film can be used on the flat-bed densitometer. Usually the resolution of the rotating drum densitometer is inferior to the flat-bed type and the micrograph may have to be photographically enlarged so that fine detail in the image is not lost in the scanning process. These densitometers must be computer controlled because they operate at speeds as fast as 20,000 data points/second and the digital information is either written onto *magnetic tape* or onto the storage facility of a computer, the *disc*. It is impractical to store the information from say a 256 × 256 image scan on paper tape, firstly because the production of paper tape is too slow and secondly because it would produce an unmanageable length of tape!

The light spot of the densitometer should be significantly smaller than the size of detail to be resolved in the electron micrograph, at least one half the size. Thus, if the extent of the optical diffraction pattern of the micrograph shows that structural detail is preserved to r nm (§ 3.3), then the image should be scanned at least at $r/2$ (nm) and preferably $r/3$ or $r/4$. For an image taken at magnification M, the scan spot should be smaller than $D = Mr/2$ (D in μm, M in units of 1000, r in nm) so that no information is lost in the numerical processing (Frank 1973). For example, if an optical diffraction pattern shows diffraction orders out to 1 nm resolution and the image magnification is 40,000, then the image should be scanned with a spot size of less than $40 \times 1/2 = 20\ \mu$m. Normally the incremental steps in the movement of the micrograph across the light spot are the same as the spot size D. For most micrographs scans are produced from square areas of micrograph, but they need not have the same dimensions in the x- and y-directions, nor need the incremental steps of the densitometer be the same in both directions. Helical particles, for example, are scanned on a rectangular raster.

Whenever possible images of crystalline specimens should be aligned so that the scan direction corresponds to the principal axes of the crystalline lattice; this is impossible for non-orthogonal axes (e.g. a hexagonal lattice) and is very difficult for low contrast images. This alignment of the image with respect to the scan is necessary because, unlike the optical transform, the numerical transform is a *discrete* or *sampled* Fourier transform, which assumes that the scanned image is exactly periodic. The consequences of using a discrete transform are examined in the following sections.

3.4.2   The fast Fourier transform (FFT)

The fast Fourier transform (FFT) is a computer program designed to calculate a numerical transform more rapidly than would be calculated from Eq. (3.5). The integrations of Eq. (3.5) (see § 3.1) are replaced by summations; this is equivalent to using a Fourier series in which an image $j(n'\Delta x, m'\Delta y)$ is sampled at equidistant intervals Δx and Δy in the x- and y-directions:

$$J(v_x, v_y) = \sum_{n'=1}^{n'=NX} \sum_{m'=1}^{m'=NY} j(n'\Delta x, m'\Delta y)\exp\left[-2\pi i(n'\Delta x v_x + m'\Delta y v_y)\right] \quad (3.38)$$

for an image scan of NX points in the x-direction and NY points in the y-direction. Normally $NX = NY$, e.g. 256×256. The spatial frequency in the transform J is replaced by the sampled frequency values (v_x, v_y) $= (n/l_x, m/l_y)$ where l_x and l_y are the scan lengths along the x- and y-directions respectively, and because of the nature of the FFT l_x and l_y define the periodicity of the image in the x- and y-directions. Only when non-equal Δx and Δy are necessary will they be introduced into Eq. (3.38); e.g. for helical structures (§ 3.6.3). So taking $NX = NY = N$ and $l_x = l_y = l$, the Fourier coefficients $J(n, m)$ of the image are given by:

$$J(n, m) = \sum_{n'=1}^{n'=N} \sum_{m'=1}^{m'=N} j(n'\Delta x, m'\Delta y)\exp\left[-2\pi i\left(\frac{n'n}{N} + \frac{m'm}{N}\right)\right] \quad (3.39)$$

since

$$n'\Delta x \cdot \frac{n}{l} = \frac{n'\Delta x n}{N\Delta x} = \frac{n'n}{N}$$

Thus the Fourier transform is displayed as a set of Fourier coefficients $J(n, m)$ indexed in terms of a set of integers (n, m) as shown in Fig. 3.19. The actual spatial frequency can always be calculated from a particular value of (n, m), since $v = \sqrt{n^2 + m^2}/l = (M\sqrt{(n^2 + m^2)})/(DN)$, but it is usual to work with a numerical transform in terms of sample points (n, m) rather than the

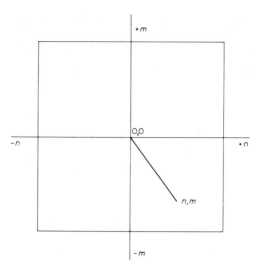

Fig. 3.19. Determining the spatial frequency from the output of the fast Fourier transform. n and m define the coordinates of a transform point.

actual spatial frequencies $(v_x, v_y) = (n/l, m/l)$. Note that the maximum values of n and m in the transform correspond to $N/2$, so that the maximum spatial frequency in the transform is $N/2l$ or $1/2\Delta x$ and the maximum resolution preserved in the transform is $2\Delta x$ or twice the scan spot size; this explains why the image must be sampled at least at $r/2$ (nm). Note that Δx, Δy and l all refer to distances in the *specimen* and to obtain the actual distances on the *micrograph* these distances should be multiplied by the magnification M of the micrograph. It is desirable to work always in terms of specimen distances independent of M or preferably in terms of transform sample points (n, m).

Provided that certain precautions are taken the FFT should have the same mathematical properties as the original Fourier transform. The following mathematical relations are needed in order to use the FFT:

(i) As the image intensity is real, its Fourier transform is *complex conjugate symmetric*, that is:

$$J(v) = J^+(-v)$$

or

$$J(n, m) = J^+(-n, -m) \qquad (3.40)$$

so that the intensities of diametrically opposite Fourier coefficients in Fig. 3.19 are equal: $|J(\mathbf{v})|^2 = |J(-\mathbf{v})|^2$. This means that it is only necessary to calculate one-half of the numerical transform because the second half can always be generated by this symmetry (also known as Friedel's law); although the intensities are the same, such symmetry related spots may have opposite phases ($-\omega°$ instead of $\omega°$). Thus from the transform intensities alone it is impossible to distinguish between, for instance, 2-fold (P2) and 4-fold (P4) symmetry or 3-fold (P3) and 6-fold (P6) symmetry of a lattice, because the diffraction intensities have built-in 2-fold symmetry.

(ii) The amplitudes of the Fourier transform are *translationally invariant*, that is, the Fourier amplitudes of an image with a coordinate system (x, y) are the same as an image with coordinates displaced by $(\delta x, \delta y)$ to $(x + \delta x, y + \delta y)$. This means that it does not matter where the origin of coordinates of the image is chosen for calculating the Fourier transform amplitudes. However, the phases of the transform do depend on the origin of coordinates; a change of origin $(\delta x, \delta y)$ causes a phase shift in the transform of:

$$-2\pi(v_x\delta x + v_y\delta y) \tag{3.41}$$

Thus if phase information is required from a numerical transform the *phase origin* must be determined. For example, if the phases of diffraction spots from a crystalline lattice are required, the phase origin of the transform corresponds in the image to the centre of one of the repeating units in the structure. So if the centre of the image is to be translated by $(\delta x, \delta y)$, the transform should be multiplied by a factor $\exp\left[-2\pi i(v_x\delta x + v_y\delta y)\right]$ or in terms of sample points $\exp\left[(-2\pi i/N)(n\delta n_x + m\delta n_y)\right]$, where δn_x and δn_y are the changes in the origin in terms of sample points n_x and n_y.

The translational invariance of the amplitudes of the Fourier transform means that by using amplitudes only it will be impossible to separate the transforms of two superimposed layers of a structure that are only translationally and not rotationally displaced (e.g. λ-bacteriophage polyheads, see § 4.3); only the phases will be affected enabling a distinction to be made between one or two layers. Rotational displacement of the two layers will, however, be observed in the transform as pairs of spots.

There are two other properties of the FFT that should be known: the origin of the image corresponds to the top left-hand corner of the scanned image, and in order to translate the origin to the centre of the image, all transform values should be multiplied by the phase factor $\exp\left[-\pi i(n + m)\right]$ corresponding to an origin shift of $(x_0, y_0) = (N\Delta x/2, N\Delta y/2)$. In the FFT the Fourier coefficients are stored in the computer as complex numbers (real and imaginary parts) in lines, starting from $n = 0$ to $n = +NX/2$, and then continuing from $n = -NX/2 + 1$ to $n = -1$. Thus for visual assessment, it is usual to reorder the coefficients to form lines from negative $n(-NX/2+1)$ through 0 to positive $n(+NX/2)$.

The second property of the discrete transform which must be considered is the assumption that the image is periodic and is sampled over an exact number of the repeat units. It is very difficult to ensure that the scan step size and the image magnification are matched to satisfy this condition; inexact sampling causes the numerical transform maxima to be blurred and its overall appearance to be inferior to the optical transform. In order to reduce the effect of non-periodicity, the scanned image is *floated* to a mean density of zero before the transform is calculated; the average density of the image is calculated and subtracted from the optical density values for each image point stored in the computer. Once the transform has been calculated it is necessary to *output* the information. Photographing the transform from a television display, where numbers have been converted to grey levels, is inadequate, because advantage should be taken of the quantitative information in the numerical transform. The most common way to output the transform is onto a *line-printer*; Fig. 3.20 shows the transform amplitudes of a micrograph of a neuraminidase crystal; they have been scaled between 0 and 100 and printed in such a way that the frequency scales horizontally (v_x) and vertically (v_y) are nearly equal, so that the symmetry of the lattice can be seen more easily. Only one half of the transform is displayed because the other half can be generated by symmetry, using $|J(n, m)| = |J(-n, -m)|$. The transform shown is from a 256×256 image scan, so it has 256×129 amplitude values. This cannot be printed in a single block on line-printer paper which has a maximum width of 130 characters; the normal format for such an output is 40I3 (40 integer nos. line, each number occupying 3 spaces) printed on alternate lines; thus the transform in Fig. 3.20 was made up from several sheets aligned and taped together. The origin (0, 0) is marked with a double circle, whilst diffraction maxima are denoted by single circles. It can be seen that the diffraction maxima are not sharp; this is partly due to disorder in the crystal and partly to inexact sampling of the lattice when the

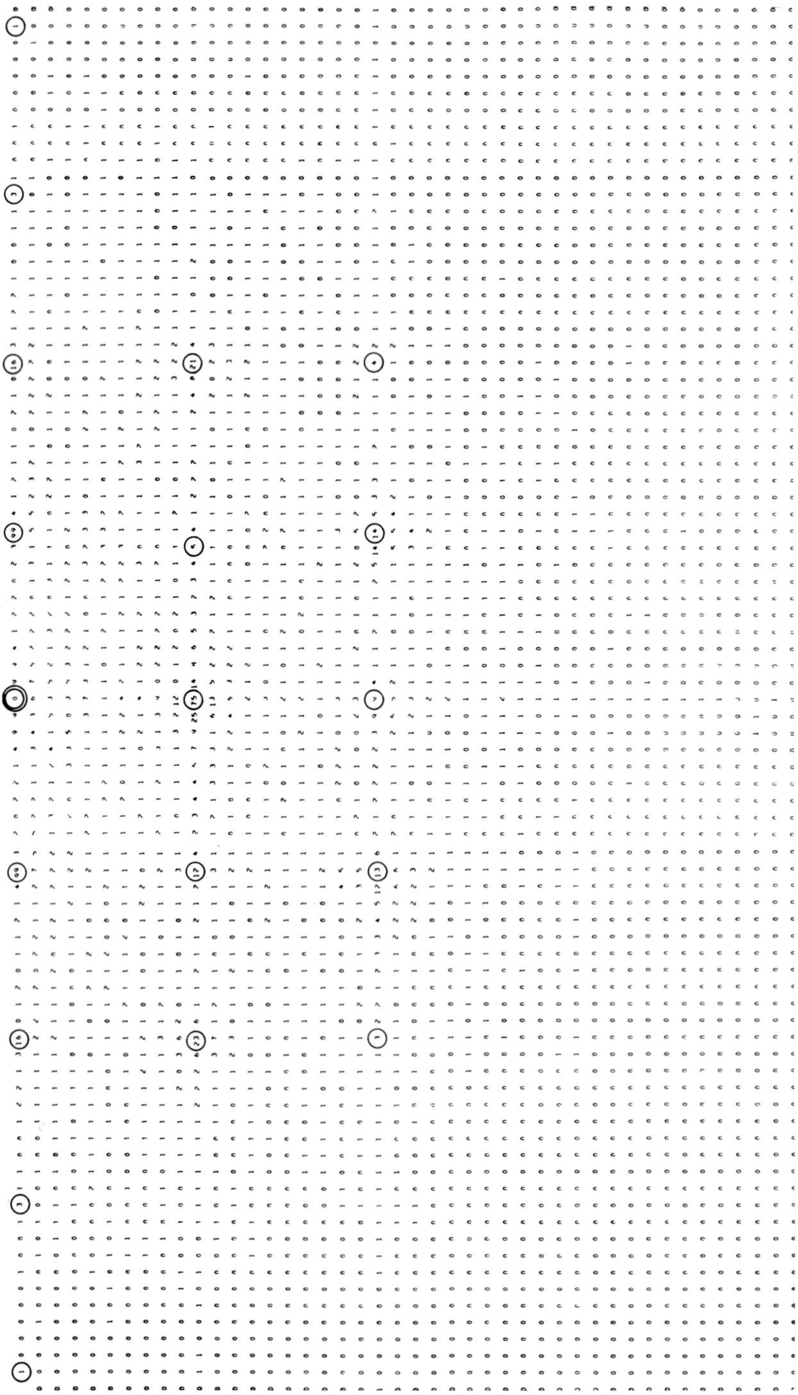

scanned image was produced. The latter effect can be seen in the difference between the two first-order maxima, the spots with amplitudes 99 and 75, which should be the same for a P4 lattice, although it can be seen that the *integrated amplitudes* obtained by adding all the numbers in the vicinity of the spot (less the background) are nearly the same. The criterion for a genuine diffraction order is that its amplitude should be clearly distinguishable from the background, which generally decreases as n and m increase. An amplitude of 4 on a background of 1 is as significant as one of 50 on a background of 10, for example. It may be that scaling amplitudes between 0 and 100 is insufficient for showing up weak diffraction orders and it is then necessary to use either a larger *dynamic range* (say 0–1000) or to output the transform in several sections with a different, known, scaling factor for each section. In practice, such low intensity diffraction orders often contribute very little to the image reconstructed from the transform (§ 4.2.3).

The spatial frequency of the diffraction maxima can be calculated from Fig. 3.20 using the information that the image ($M = 52{,}000$) was scanned in incremental steps of 20 μm ($\equiv 0.38$ nm on the specimen) on a 256×256 raster; so $v = (M/DN)\sqrt{(n^2 + m^2)} = 0.0102\sqrt{(n^2 + m^2)}\,\mathrm{nm}^{-1}$; M is in units of 1,000 and D is in μm. The spacing of the transform maxima in both the 'x' and 'y' directions is almost exactly 10 sample points so that the reciprocal lattice constants are both $0.102\,\mathrm{nm}^{-1}$; thus $a = b = 1/0.102 = 9.8$ nm. The resolution of the image can be determined from the highest values of (n, m) giving a diffraction maximum; in Fig. 3.20 this corresponds to $\sqrt{(n^2 + m^2)} = \sqrt{(30^2 + 0^2)} = 30$ or $v_{\max} = 0.31\,\mathrm{nm}^{-1}$ (3.2 nm resolution). Optical diffractometry would have produced this information more easily, but as well as the amplitudes the phase information can be displayed separately; normally the phase angles (between $0°$ and $360°$) are scaled between 0 and 36 and displayed in the same way as the amplitudes. The positions of the intensity maxima are then marked on the phase output, or the information from both the amplitude and phase maps are marked on a single piece of graph paper with say 1 mm equal to 1 transform sample point ($\Delta n = 1$).

Fig. 3.20. Part of a line-printer amplitude display of a numerical transform. The origin is indicated by a double circle and transform maxima are circled. The transform was calculated from an image of a negatively-stained (1 % uranyl acetate, UA) neuraminidase crystal at magnification $M = 52{,}000$. The spot ($\pm 30{,}0$), the highest order of any significance, corresponds to a resolution of 3.2 nm.

3.4.3   *The analysis of numerical transforms: amplitude and phase information*

The best way to illustrate the use of numerical transform data is by reference to specific applications:

Example 1   Focus series of images have been obtained from unstained ribosome crystals (Unwin 1976, unpublished). The image transforms need to be examined in detail to ascertain whether the weak phase approximation is valid for these images, before proceeding with image reconstruction. Additionally where image transforms show frequency gaps these should be filled in by using image transforms with frequency gaps in different positions.

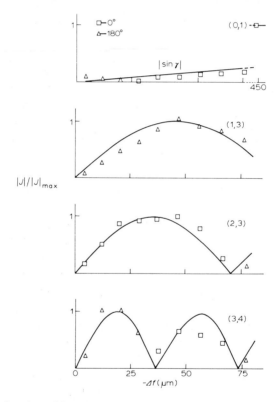

Fig. 3.21.   Testing the validity of the weak phase approximation for a crystalline array of ribosomes. The moduli of the Fourier components $|J|$ are normalised by $|J|_{max}$ and plotted against $\sin\gamma$ for a range of Δf. Note the large value of Δf required for phase contrast. (0,1), (1,3), (2,3) and (3,4) are the indices (h, k) for particular diffraction maxima. ($\square$) and ($\triangle$) indicate the phase of the Fourier components $|J|/|J|_{max}$ (P.N.T. Unwin, unpublished.)

The double-layered ribosome crystals have a P422 lattice ($a = 59.5$ nm) and are about 60 nm thick, rather thicker than one would expect for the validity of the weak phase approximation.

In order to produce phase contrast images for such large spacings, defocus values in the range 20–500 μm must be used (§ 3.2). Figure 3.21 shows the transforms for particular diffraction maxima $[(0, 1), (1, 3), (2, 3)$ and $(3, 4)]$. For each diffraction order, the ratio of its amplitude $|J|$ to the maximum transform amplitude of the whole series $|J|_{max}$, is plotted against the image defocus Δf (Unwin, personal communication). The curves represent the envelope of $\sin \gamma$. The fact that the amplitudes fit the $\sin \gamma$ curve quite well indicates that the weak phase approximation ($|J|$ proportional to $|A| \, |\sin \gamma|$) is valid. The maxima in $|J|$ occur very nearly at the values for Δf that would be predicted using $\sin \gamma = 1$ for maximum phase contrast: for a spacing $r = 1/v$, maximum phase contrast occurs when $\Delta f = r^2/2\lambda$ ($\lambda = 3.7$ pm). For the ribosome lattice $v = a^* \sqrt{(h^2 + k^2)}$, with $a^* = 1/59.5$ nm^{-1}, so that for the $(0, 1)$ diffraction maximum, $r = 59.5$ nm, $\Delta f = 478$ μm; for $(1, 3)$, $r = 18.8$ nm, $\Delta f = 47.8$ μm; for $(2, 3)$, $r = 16.5$ nm, $\Delta f = 36.8$ μm; and for $(3, 4)$, $r = 11.9$ nm, $\Delta f = 19.1$ μm. So it is with some confidence that the images represented in Fig. 3.21 can be used for reconstruction based on the weak phase approximation. It is not easy to obtain a focus series such as shown in Fig. 3.21 because ideally the images should all have the same mean optical density so that transform values from different images can be compared or even combined to form a composite transform that has no transfer frequency gaps, and includes only those transform points where $\sin \gamma$ is near to unity or at least significantly larger than the background noise.

Example 2 shows the use of amplitude and phase information in the transform to give some indication of the symmetry of a crystalline lattice and some idea of how the molecules are arranged in the crystal. Figure 3.22 shows the results obtained from a numerical transform of the image of a 'square' neuraminidase crystal, where diffraction maxima with approximately equal amplitudes have been given the same number. The transform clearly shows 4-fold symmetry by the near equality of spots related by a 90° rotation. In addition, diffraction orders on either side of two lines drawn at 45° to the principal axes also have equal intensities ($4 = 6$, $8 = 12$, $9 = 11$), which suggests another axis of two-fold symmetry, a mirror plane. Because of the 4-fold symmetry this generates at least 2 more mirror planes so that the space group for the neuraminidase crystal is most likely to be P422 (International Tables for X-ray Crystallography, Vol. I), but little more can be said about

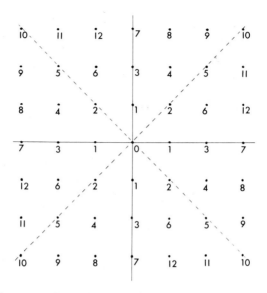

Fig. 3.22. The image transform of a 'square' neuraminidase crystal. Transform maxima marked with the same numbers indicate 4-fold symmetry of the lattice. Additional 2-fold symmetry (mirror) planes are indicated by the equal intensities and phases of pairs of diffraction maxima such as 4 and 6, 8 and 12, 9 and 11 (see Table 3.3).

the molecular arrangement within the unit cell without information on the third dimension*. However, the phase information can be used to confirm

---

* Sometimes it is possible to obtain some indication of the c-axis dimension without taking a tilt series, by using the relationship between the volume V of the unit cell, the number of molecules in the unit cell, n_0, the density of the crystal ρ and the molecular weight MW of the molecules forming the crystal:

$$\text{Mass of one molecule } V\rho = \frac{n_0 MW}{N_0} \tag{3.42}$$

where MW in kg contains Avogadro's number (N_0) of molecules. If the units of V are nm^3 and the units of ρ are $kg\ m^{-3}$, Eq. (3.42) becomes:

$$V = \frac{n_0 MW}{\rho \times 6.023 \times 10^{-1}} \tag{3.43}$$

For neuraminidase, the molecule is a tetramer of dimensions $8 \times 8 \times 4$ nm with a molecular weight of about 200,000 (Wrigley et al. 1973); MW should include an estimate for the normal water content of a protein crystal (about 25 %), that is $MW = 270,000$ – this does not necessarily mean that the water content is retained in the negatively-stained specimen but makes an allowance for the 'open spaces' in the crystalline lattice. For protein crystals $\rho = 1250\ kg\ m^{-3}$. For

P422 symmetry. All phases should be 0 or π and additionally phases of spots related by 4-fold symmetry should be identical (International Tables for X-ray Crystallography, Vol. I). Before the phases can be tabulated, the phase origin must be determined (§ 3.4.2); in this example this involves determining the origin shift (δx, δy) or (δn_x, δn_y) in order to give phases as near to 0 or π as possible or, alternatively make 4-fold related phases equal. This can be done by eye by examining pairs of diffraction phases and calculating (δn_x, δn_y) so as to make them equal; this is repeated for several pairs of spots, noting that every pair should give a consistent value for (δn_x, δn_y). For a diffraction maximum with coordinates (n, m) the phase shift caused by an origin shift (δn_x, δn_y) is $\delta\omega = (-2\pi/N)(n\delta n_x + m\delta n_y)$ radians. This same origin shift causes a phase shift of $(2\pi/N)(n\delta n_x + m\delta n_y)$ in the diffraction maximum with coordinates ($-n$, $-m$). Thus the total phase shift caused in this pair of maxima by altering the origin by (δn_x, δn_y) is $\Delta\omega = 2\delta\omega$. For simplicity consider $m = 0$, when the total phase shift between a pair of maxima $-n$ and $+n$ is $\Delta\omega = (4\pi/N)(n\delta n_x)$ or $\delta n_x = (\Delta\omega N/4\pi n)$, or for $\Delta\omega$ in degrees $\delta n_x = (\Delta\omega/720°)(N/n)$. For example, the two spots with $n = -10$ and $n = +10$ (the $(-1,0)$ and $(1,0)$ diffraction orders) (Fig. 3.20) initially had phases of 0° and 20°, respectively. To make the two phases equal required respectively the addition and subtraction of 10° so the total phase shift $\Delta\omega = 20°$ = difference in the two phase values. Thus $\delta n_x = (20/720)$ $\times (256/10) = (256/360) < 1$ sample point. Several independent values for δn_x can be obtained for $m = 0$. Similarly for diffraction orders with $n = 0$, values for δn_y can be determined using $(0, m)$ maxima. It is then possible to check that the phases of other diffraction orders (n, m) can be symmetrised using a single value of (δn_x, δn_y). This is tedious with a large number of diffraction spots, and it may be more convenient to use a computer program which tries several values of δn_x and δn_y to find the best set of symmetrised phases. Table 3.3 shows a set of amplitudes and phases for the neuraminidase crystal transform shown in Fig. 3.22. Despite the determination of a phase

neuraminidase, from the symmetry of the lattice, the number of molecules in the unit cell, n_0, should be a multiple of four, and since $V = a \times b \times c$ for an orthorhombic crystal:

$$c = \frac{4 \times 2.7 \times 10^5}{9.8 \times 9.8 \times 1250 \times 6.023 \times 10^{-1}} = 14.9 \text{ nm}$$

with $a = b = 9.8$ nm. This is only an estimate of c since the water content of protein crystals may be as high as 50% ($MW = 400,000$). It is more usual to use Eq. (3.42) to calculate n_0, knowing the dimensions of the unit cell, a, b and c. c must be checked by other methods, such as tilting and X-ray crystallography.

TABLE 3.3

Amplitudes and phases from the numerical transform of a micrograph of a negatively-stained neuraminidase crystal (see Fig. 3.22)

Number of spot (see Fig. 3.22)	h, k	Amplitude	Phase (degrees)	Phase (0 or π)
1	1, 0	99	10	0
	0, −1	95	10	0
	−1, 0	99	350	0
	0, 1	95	10	0
2	1, −1	9	190	π
	−1, −1	15	190	π
	−1, 1	9	190	π
	1, 1	15	190	π
3	2, 0	18	190	π
	0, −2	22	190	π
	−2, 0	18	170	π
	0, 2	22	190	π
4	2, −1	31	200	π
	−1, −2	25	160	π
	−2, 1	31	200	π
	1, 2	25	200	π
5	2, −2	9	200	π
	−2, −2	8	10	0
	−2, 2	9	200	π
	2, 2	8	10	0
6	1, −2	34	200	π
	−2, −1	35	180	π
	−1, 2	34	200	π
	2, 1	35	180	π

origin there are still spots where the phases differ substantially. Complete symmetrisation involves setting all symmetry-related diffraction orders to the same amplitude and the phases to exactly 0 or π. Symmetrisations should be made with caution because the reconstructed image derived from such a transform will always show the symmetry imposed upon it. It is not necessary to actually move the origin of the image; it is merely necessary to make the appropriate phase shift in the transform by multiplying all transform values by $\exp\left[(-2\pi i/N)(n\delta n_x + m\delta n_y)\right]$, once δn_x and δn_y have been determined. Although the phase origin refinement may not be essential for crystalline specimens, it is an essential part of rotational filtering (§ 4.1), where the rotational centre of the particle must be determined with precision.

3.5 Optical versus computer analysis of electron micrographs

The optical and numerical transforms should be equivalent provided that the image is scanned with the correct sampling; i.e. oversampling using a smaller scan spot and smaller incremental steps on the densitometer than $Mr/2$ (sometimes referred to as *Nyquist sampling*).

 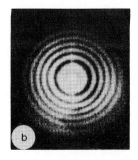 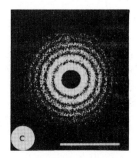

Fig. 3.23. Optical (b) and computer (c) transforms of the image (a) of a carbon film about 300 nm underfocus. The centre of the computer transform has been removed for display. Image bar = 20 nm, diffraction bar = 3 nm^{-1}. (N.G. Wrigley, unpublished.)

Figure 3.23 shows a comparison of (b) an optical transform and (c) a numerical transform of a micrograph of a carbon film (a), taken about 300 nm underfocus. The numerical transform has been displayed on a television system and its central region omitted so that the ring pattern can be seen clearly. With amorphous specimens like this, displaying the numerical transform on the line-printer is visually unsatisfactory and it is very difficult, and inaccurate, to try to measure the positions of minima on a line-printer display.

For ease of use, the optical diffractometer is preferable to numerical methods. But there are two fundamental disadvantages in using optical diffractometry for quantitative evaluation of image transforms; these are the problems of displaying transform data on a linear intensity scale and the loss of phase information. In a computer calculated transform, it is relatively easy to measure the relative intensities of the diffraction spots and to decide which diffraction spots are significantly above the background level. And if image reconstruction is to be applied to a set of images, it is relatively easy to combine numerical data but optical combination of images is very difficult. During image reconstruction, the transform can be numerically symmetrised; this is an allowed procedure provided some confidence can be placed in the indexation of the lattice (§ 3.7). All one is doing when symmetrising a transform is to put back information that was lost as a result of the damage and distortion of the specimen caused by the preparation methods and radiation damage. Superimposing incorrect symmetry on a transform can lead to a false structure but provided it has been analysed carefully (§ 3.4) and indexed using a single reciprocal lattice construction (§ 3.7), there should be no doubt about the validity of a computer reconstruction. Optically, symmetrisation can be achieved by photographic superposition (§ 4.1 and § 4.2) but computer symmetrisation is less tedious. Optical and numerical transforms should be considered to be complementary; whereas a large number of images can be screened by optical analysis, a few of the 'best' should be selected for computer analysis, followed by image reconstruction.

The most worrying feature about the numerical transform is the loss in quality of the diffraction orders as a result of inexact sampling. For a non-crystalline specimen the numerical and optical transforms are of similar quality, provided that the numerical data has been floated to a mean density of zero. However, the numerical transform of a crystalline lattice is often inferior to the optical transform, with an apparent loss of high resolution information. Assuming that the same area of micrograph is used in both systems, quality is lost in the numerical transform due to the blurring of the diffraction orders unless three conditions are satisfied:

(i)   There should be an exact number of sample points in a single repeat unit of the lattice;

(ii) There should be an exact number of repeat units in the scanned area;

(iii) The scan should be aligned with the principal axes of the crystal lattice.

All these conditions are difficult to satisfy with noisy images, and impossible to satisfy for crystalline specimens with non-orthogonal axes (e.g. a hexagonal lattice).

A comparison between the optical and numerical transforms from an image of two overlapping T-layers from *Bacillus brevis* is shown in Fig. 3.24 (Aebi et al. 1973); the transform is taken from the image area A. The optical transform (left-hand half of Fig. 3.24b) shows two tetragonal lattices rotated with respect to each other, corresponding to the two layers of the cylinder. The optical transform is clearer than the numerical transform (right-hand half of Fig. 3.24b), which was displayed using overprinting of characters on a line-printer to produce intensity variations. It is impossible to find a sampling of the image that will be optimum for both layers, so the numerical transform spots generally appear to be less well defined than in the optical

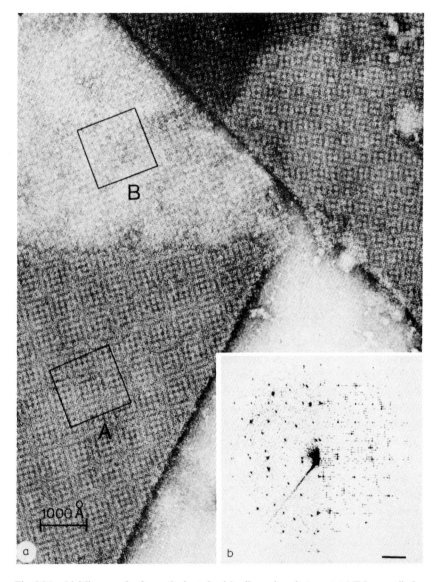

Fig. 3.24. (a) Micrograph of negatively-stained (sodium phosphotungstate) T-layer cylinder, (b) diffraction pattern of the double layer marked A: left-hand side is an optical transform; right-hand side is a computer transform. Image bar = 100 nm, diffraction bar = 0.1 nm^{-1}. (From Aebi et al. 1973.)

transform. It is worth noting that although area B in Fig. 3.24a corresponds to a single T-layer, its transform shows that it is not as well ordered as the double layer at A. Many biological systems do occur naturally as 'cylinders' (bacterial cells, microtubules) which flatten on the specimen grid to produce superimposed lattices and often this double layer is a more stable structure than one obtained by attempting to 'open' or break the structure into a single layer, although multilayered structures can create more complicated diffraction patterns, and consequently uninterpretable images.

For an image that is not *correctly sampled* due to an incorrect choice of scanning parameters it is possible to do a *coarse filtering* first (§ 4.2) to remove some of the image noise. From this image, it is usually possible to work out a new coordinate system that satisfies the three conditions given above for optimum sampling of the lattice and then to interpolate the original image density variations onto this new lattice (§ 4.2). Small distortions in the lattice can also be corrected by these means (Crowther and Sleytr 1977); such distortion cannot be corrected by optical means.

If only image transforms are to be analysed then it is doubtful whether it is necessary to go to the trouble of densitometry and computer analysis. But if image processing is contemplated then the numerical methods have several advantages over optical image reconstruction methods, such as the ease of symmetrising structural information, combining images and correcting for the effects of the transfer function $\sin \gamma$. Also, expensive optics have to be bought for optical bench reconstruction, whereas many establishments already have computing facilities at their disposal. Optical equipment can also be sensitive to external vibrations, so special arrangements for their installation have to be made.

In the reconstruction of helical particles, numerical processing is able to provide a three-dimensional reconstruction from essentially a two-dimensional transform; optical filtering is only capable of producing a 'one-sided' image of the helix (§ 3.6). For the crystallographer the access to phase information via a numerical transform will be important. It is well known that in X-ray crystallography the phase information in the transform is more important than the amplitude information (Sherwood 1976); knowing the amplitudes of the transform $|A|$ alone gives only the auto-correlation function of the 'structure' η (the *Patterson function* in X-ray crystallography). In the electron microscopy of thin specimens (that is, those specimens satisfying the weak phase approximation) it is possible to obtain both amplitude and phase information, $|A|$ and ω, and based on crystallographic studies advantage should be taken of this knowledge. In the classic paper on the electron microscopy of unstained biological specimens by Unwin and Henderson (1975), accurate and symmetrised phases are obtained from a numerical image transform, whilst accurate amplitudes $|A|$ (unaffected by the transfer function $\sin \gamma$) are obtained from electron diffraction data (§ 4.5). Numerically it is relatively easy to combine ω and $|A|$ to produce a specimen transform $|A| \exp (i\omega)$, whereas optically this procedure would be difficult to implement.

In conclusion optical transforms are an excellent means of selecting images with good preservation of structure and an acceptable transfer function. Selected images can then be further analysed using numerical transforms.

3.6   Diffraction by helices

Helical symmetry occurs in several biological complexes such as the tails of T-even bacteriophages, tobacco mosaic virus, intact microtubules and some nucleoprotein complexes. The protein subunits in these particles are arranged on a helix so that a single projection (side-on view) gives many different views of the same molecule (§ 2.3). So, a two-dimensional projection, looking side-on to the helix, contains three-dimensional information on the structure, subject to the resolution limit d, imposed by the number of subunits N in the axial repeat distance of the helix (i.e. the number of different views): $d = \pi D/N$ (Eq. (2.2)), where D is the diameter of the subunits in the helix. The main characteristic of a helix is its *pitch P* which corresponds to the axial distance along the helix that gives a rotation of 2π (360°) on its surface. So if the separation p (approximately equivalent to D) between subunits is not much smaller than P, the number of independent views of the subunit $N = P/p$ in a complete helical repeat will not be large, and a three-dimensional reconstruction will give only a low resolution result ($d \simeq \pi p^2/P$). Solving a helical structure from its diffraction pattern requires the determination of the parameters p and P (Fig. 3.25), or p and Ω, the *screw-angle* of the helix, the angular displacement of successive structural units ($\Omega = 2\pi p/P$).

A helix could also be considered to arise from rolling a flat sheet of subunits into a cylinder in such a way that the subunits follow a helical path on the surface of the cylinder. This leads to the idea of considering the optical diffraction pattern of a helix to arise from the superposition of patterns from the images of the top and bottom surfaces of the cylinder. This means that it is possible to find two independent lattices and make a reconstruction from just the bottom or top surface, as is normally made for specimens comprising superimposed layers such as T-layers (Aebi et al. 1973) and

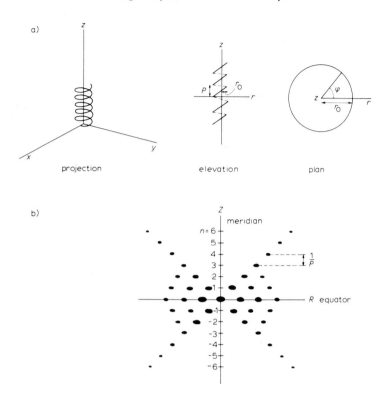

Fig. 3.25.   Diffraction by helices: (a) a continuous helix, its projection and the definition of polar coordinates (r, ϕ). P = pitch of the helix. (b) the diffraction pattern with a layer line spacing of $1/P$. (From Sherwood 1976.)

polyheads from T4 phages (Steven et al. 1976a and b). However, a helical particle is often a solid three-dimensional object (e.g. TMV) supported by negative stain, not an empty cylinder, and the projection of one-half of the helix is not much more informative than the projection of the whole helix. In some cases when a helical structure comprises a well-preserved thin walled tube (e.g. a microtubule) the selection of two sets of diffraction spots can clearly separate the top and bottom of the structure, except at the edges of the particle, which are highly curved (Klug and DeRosier 1966).

The basic theory of diffraction by helices (Cochran et al. 1952; Klug et al. 1958) will now be considered, and it will be shown that it is possible to identify helical symmetry from image transforms and to attempt to determine the helical parameters which indicate how the subunits are arranged on the helix.

3.6.1 The diffraction pattern of a continuous helix

A continuous helix will serve as a model to indicate the characteristics of a helical diffraction pattern before considering the more complicated, but real case of a discontinuous helix composed of discrete subunits. Figure 3.25a shows the projection of a left-handed helix for which a counter-clockwise rotation is accompanied by a translation in the $+z$-direction. The electron microscope image normally gives a side-on view (elevation) of a helix with an axial (z) repeat P and a radius r_0 as shown in Fig. 3.25a. It is normal to define a helix in terms of the cylindrical polar coordinates (r, ϕ, z) (Fig. 3.25a). Mathematically the density variations ρ in the helix satisfy the relation:

$$\rho(r, \phi, z) = \rho(r, \phi + 2\pi, z + P) \tag{3.44}$$

where P is effectively the repeat distance of the helix in the z-direction; the helix is essentially a one-dimensional lattice in the z-direction, so its transform is expected to be discrete in the z-direction, but continuous in all other directions. Thus the Fourier transform of $\rho(r, \phi, z)$, $F(R, \Phi, Z)$, will have an axial repeat along the Z-axis (the *meridian*) of $1/P$. This generates a set of equally spaced *layer planes*, or, in two-dimensions, *layer lines*, separated by $1/P$ and indexed from $Z = 0$ (the *equator*) $n = 0, 1, 2 \ldots$ in the Z-direction and by negative integers in the $-Z$-direction, as shown in Fig. 3.25b. The amplitudes $|F|$ of the diffraction orders are proportional to the Bessel functions of order n, $J_n(2\pi R r_0)$ (Cochran et al. 1952; DeRosier and Moore 1970), where r_0 is the radius of the particle and R is the reciprocal radial distance. It is sufficient to know that Bessel functions, J_n, are rather like damped cosine or sine waves as shown in Fig. 3.26 and they are tabulated just as cosine and sine functions (Abramowitz and Stegun 1965). Thus along a particular layer line the transform intensities are proportional to $[J_n(2\pi R r_0)]^2$ so the lines become broken and like a set of spots. Because only J_0 is non-zero for $R = 0$, only the $n = 0$ layer line will be non-zero on the meridian. Successive first maxima of J_n progressively occur at larger values of R (Fig. 3.26). The diffraction maxima for increasing n will occur further from the Z-axis (Fig. 3.25b) giving the appearance of a cross with the intensity decreasing as n increases. This cross is the characteristic appearance of the transform of the image of a helical particle. Because the maxima of J_n are broad, the diffraction spots will be elongated in the R direction.

Because of the following relations:

$$J_{|n|}(X) = (-1)^{|n|} J_{|n|}(-X)$$

$$J_{-|n|}(X) = (-1)^{|n|} J_{|n|}(X) \tag{3.45}$$

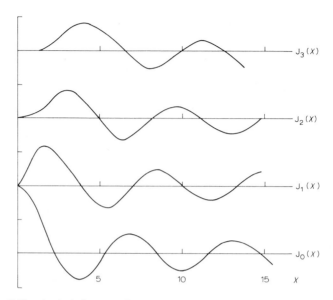

Fig. 3.26.   Diffraction by helices: amplitude variations of the Bessel function $J_n(X)$ for $n = 0$ to 3. The progressive displacement of the first maxima explains the cross shape of the diffraction pattern in Fig. 3.25b.

the intensities of diffraction orders with the same value of n but on opposite sides of the meridian or equator (i.e. n positive or negative) should be equal. Thus an examination of the transform of the image of a negatively-stained helical particle enables the uniformity of staining and any distortions to be assessed from the differences in intensity between such a set of four diffraction maxima. It should be noted that the diffraction intensities $|F|^2$ are independent of the angular coordinate Φ and depend only on $n(Z)$ and R. However, the phases of F do depend on Φ.

3.6.2   The diffraction pattern of a discontinuous helix

Biological helices are not continuous distributions of density but are a series of identical subunits separated in the z-direction by a distance p on a helix of pitch P as shown in Fig. 3.27a; these subunits are successively rotated through the screw angle Ω (§ 3.6.1). Initially it will be assumed that there is an exact number of subunits in a distance P, that is, P/p is an integer (in Fig. 3.27a, $P/p = 5$); each subunit is rotated by $\Omega = 2\pi p/P$ with respect to the one below it. The discontinuous helix can be represented as the *multiplication* of a continuous helix by a one-dimensional lattice with a repeat distance p

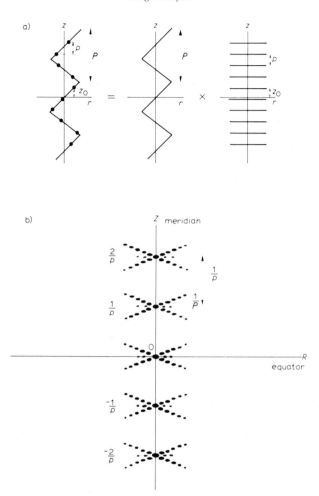

Fig. 3.27. Diffraction by helices: (a) a discontinuous helix, with subunits spaced in z by p and the helix pitch P, is equal to the multiplication of a continuous helix by a lattice with repeat distance p, (b) the diffraction pattern, with layer lines separated by $1/P$ and successive meridional diffraction orders separated by $1/p$. (From Sherwood 1976.)

(Sherwood 1976). Thus the Fourier transform of such a helix is a convolution of the transform of a helix with a one-dimensional lattice characterised by a reciprocal lattice constant $1/p$. The diffraction pattern of a discontinuous helix thus consists of a set of meridional orders separated by $1/p$ and indexed by the integer m. Centred at each particular value of m is a diffraction pattern arising from the helical nature of the particle. As shown in Fig. 3.27b, the diffraction pattern consists of a series of crosses with a layer line spacing of $1/P$, and a spacing of $1/p$ between successive meridionals; each *branch* of the diffraction pattern corresponds to a different value of m and within each branch occur layer lines corresponding to different n. In Fig. 3.27b the separation of different branches has been exaggerated to illustrate the difference between p and P. In practice, P may not be much larger than p so that successive branches will overlap to some extent and each layer line will contain contributions from other values of n. Fortunately these additional contributions will mainly arise from higher orders of J_n which will not only be further out from the meridian but will also be of lower intensity. Success in indexing and interpreting the transform of a helix depends on being able to identify the n values and the parameters of the helix. Optical diffraction patterns showing meridional spots can be used to calculate the subunit separation p and the layer line spacing determines P. Normally there is not an exact number of subunits in P; that is, successive subunits are not rotated by a multiple of $2\pi p/P$. If the actual repeat unit of the helix is c, the layer line separation is not $1/P$ but $1/c$, and there should be only non-zero intensities corresponding to the layer line *altitude* $Z = l/c$, where l is an integer. The altitude is defined as the vertical distance Z of a layer line from the equator. For a simple helix only certain J_n can contribute to each layer line, corresponding to the *selection rule* (Klug et al. 1958; Moody 1967):

$$\frac{l}{c} = \frac{n}{P} + \frac{m}{p} \qquad (3.46)$$

If $c = P$ there is no restriction on n, but if, for example, there are 5 1/3 subunits in P ($5\ 1/3 p = P$), $c = 3P$; then for $m = 0$, $l = 3n$, so that the main contributions to layer lines 0, 3, 6, 9, 12 and 15 arise from J_0, J_1, J_2, J_3, J_4 and J_5, respectively. Layer lines 1, 4, 7, 10, 13 and 16 contain contributions mainly from the $m = 1$ ($l = 3n + 16$) branch, J_{-5}, J_{-4}, J_{-3}, J_{-2}, J_{-1} and J_0; whilst the remaining layer lines 2, 5, 8, 11 and 14 will contain contributions from the $m = -1$ ($l = 3n - 16$) branch, J_6, J_7, J_8, J_9 and J_{10}. Because high order Bessel functions have a very low magnitude near the meridian, the intensity contributions to low values of l from high values of m will be very small. In fact in the example above only the layer lines with contributions from low values of n (say 2) will be visually evident in the diffraction pattern ($l = 0, 3, 6, 10, 13, 16$). The separation of the meridional spots ($l = 0$ and $l = 16$) will give p whilst the separation of near meridional spots ($l = 3$ and $l = 6$) will give a value for P. Although the selection rule (3.46) is useful to describe the appearance of the diffraction pattern of a helix, it is usually easier to extract the helical parameters using the equation:

$$Z = \frac{1}{p}\left(\frac{\Omega n}{2\pi} + m\right) \tag{3.47}$$

for the layer line with altitude Z. Note that the sign of Ω will determine whether the helix is left-handed (negative) or right-handed (positive). The problem of indexing the diffraction pattern reduces to fitting Eq. (3.47) to the measured layer line altitudes (Smith et al. 1976; § 3.7).

There may not be a single basic helix through all equivalent subunits, but a number of such helices, r, related by an r-fold rotational axis. Here, only Bessel functions whose orders are multiples of r will occur in the diffraction pattern. For example, an uncontracted T4 phage tail consists of annuli of six subunits each arranged on a basic helix (DeRosier and Klug 1968). The separation of the annuli p is 4.1 nm and there are approximately 7 annuli in two turns of the helix. So there are six basic helices each with a screw angle $\Omega = -4\pi/7 = -103°$, whilst the rotational separation of each subunit is $103°/6 \simeq 17°$. The layer line separation $l/c = 1/28.7$ nm^{-1} and only Bessel functions or *angular harmonics* J_{6n} contribute to the intensities on the layer lines. Figure 3.28 shows the image of a negatively-stained T4 phage tail together with one half its optical diffraction pattern (Fig.

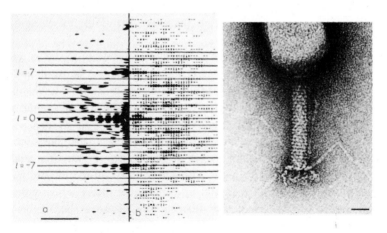

Fig. 3.28. (a) Optical, (b) computer transforms of a negatively-stained T4 phage tail (uncontracted) together with indexation of the layer lines. The spacing from the equator ($l = 0$) to the first meridional ($l = \pm 7$) = $1/4.1$ nm^{-1} and the layer line spacing = $1/(7 \times 4.1) = 1/28.7$ nm^{-1}. Image bar = 20 nm, diffraction bar = 0.2 nm^{-1}. (From Smith et al. 1976.)

3.28a) and the corresponding numerical transform amplitudes (Fig. 3.28b) with the layer lines marked approximately satisfying the selection rule $l = -2n + 7m$ (Smith et al. 1976).

If a three-dimensional reconstruction of the helix is required, it is necessary to calculate the numerical transform F in order to obtain the phase information as well as amplitude information. For the uncontracted T4 phage tail there are 42 subunits in a repeat of the helix (28.7 nm) giving 21 independent views. The ultimate resolution of such a three-dimensional reconstruction is, therefore, $d \simeq \pi \times 4.1/21 = 0.6$ nm (see § 7.4). The resolution of the reconstruction also depends on whether the Bessel functions on the layer lines overlap (DeRosier and Moore 1970).

3.6.3   The calculation of the numerical transform of a helical particle

Images of helical particles suitable for numerical analysis are selected by optical diffractometry; the particle is usually masked with opaque tape on the non-emulsion side of the micrograph to reduce the background noise in the optical diffraction pattern. Particles are selected that satisfy the following conditions:

(i)   The particles are straight, not bent;
(ii)   The layer lines are straight;
(iii)   Amplitudes of pairs of spots equidistant from the meridian are similar, indicating uniformity of negative staining. Flattened particles often give clearer diffraction patterns but these are unsuitable for three-dimensional reconstruction.

The optical diffraction pattern is indexed in terms of n and m and the parameters of the helix determined (§ 3.7); n values can be estimated from the maxima of $J_n(2\pi R r_0)$ using the diffractometer constant C to calculate R and estimating the diameter of the particle $2r_0$ from the micrograph. The selection rule may reduce the number of possible values for n, since n values for different layer lines must satisfy the same selection rule. The numerical transform will also give the phases of the transform maxima and this information can be used to assist with indexation: off-meridional pairs of spots with even values for n should have the same phase, whilst pairs with odd values of n should have phases differing by π. However, before this phase information can be used, the axis of the densitometer scan must be correctly aligned with the axis of the helix; this corresponds to determining the phase origin for the numerical transform. Also the sampling in transform space must be small enough to give adequate sampling of the layer lines, otherwise the numerical transform will be poor. The image should be scanned well above Nyquist sampling; for example, the T4 tail should be scanned with a scan step size equivalent to 0.41 nm for the axial repeat $p = 4.1$ nm. Because the helical particle is rectangular the scan dimensions are $NZ \times NR$, where NZ corresponds to the number of sample points along the z-axis (helical axis). Usually NZ is much larger than NR, although the scan step Δz is usually the same in both directions. As before, the actual scan step D (μm) on the micrograph corresponds to $M \times \Delta z$, with M in units of 1,000 and Δz in nm. The scan in the z-direction should correspond as nearly as possible to an integral number of helical repeats $c = 7 \times 4.1 = 28.7$ nm. An integral number of scan steps in each distance p and an integral number of helical repeat units c will produce an ideally sampled transform. Thus the transform sampling $\Delta Z = 1/NZ\Delta z$ should be a multiple of $(1/28.7)$ nm^{-1}. If, for example, ΔZ is chosen to be $1/57.4$ nm^{-1} then with $\Delta z = 0.41$ nm (an exact fraction of p), $NZ = 140$ scan points, whilst with the same scan step NR would be chosen to adequately cover the width (together with some background) of the helical particle; $NR = 64$ would be adequate for the width (about 20 nm) of the T4 particle. The sampling in the 'r-direction' (at right-angles to the z-axis) is unimportant because the transform is continuous in this direction; it is the sampling in the 'periodic' direction that must be exactly correct. One way to produce a micrograph that will be correctly sampled is to make a photographic enlargement of the original so that the scan step D (μm) corresponds exactly to a specimen sampling of 0.41 nm (for T4 tail). Alternatively, the original image can be scanned and then interpolated numerically so that the spacing between data points is as required (§ 4.2.3). Note that because of non-equal values for NZ and NR the sampling of the transform in the R-direction is different from that in the Z-direction; $\Delta R = 1/NR\Delta z$. Thus layer line altitudes are calculated from $Z = n_Z\Delta Z$ and positions of maxima of the J_n are calculated from $R = n_R\Delta R$, where n_Z and n_R are integers denoting the number of transform sample points in the Z- and R-directions, respectively.

After floating the image to a mean density of zero, the FFT may be calculated and displayed in the same way as transforms of other specimens, with amplitudes scaled between 0 and 100 and phases between 0 and 36 (0°–360° range). Some FFT programs will only operate on a data array with dimensions that are a power of 2 (2^n). If such an FFT is used, the scanned image after floating must be *padded out* with zeros to bring the image dimensions up to a power of 2. DeRosier and Moore (1970) favour padding out to a dimension as large as $NZ = 512$, when the transform sampling is correspondingly fine ($= 1/NZ\Delta z$) irrespective of the original scan dimensions.

Since the axis of the scan is unlikely to be exactly along the helical axis, the phases of pairs of off-meridional spots with the same value of Z cannot be compared until the phase origin corresponds to the axis of the helix. The phases ω of two transform maxima corresponding to $-R$ and $+R$ on the same layer line Z (with the same value for n) should differ by either 0 or a multiple of 2π if n is even, and only by π or an odd multiple of π if n is odd, that is:

$$\omega(-R, Z) - \omega(+R, Z) = n\pi \qquad (3.48)$$

Thus a phase difference $\delta\omega(Z) = \omega(-R, Z) - \omega(+R, Z) - n\pi$ can be compensated by an axial shift $\delta r = (\delta\omega/4\pi)(1/R)$ or in terms of image sample points, $\delta n_r = (\delta\omega/4\pi)(NR/n_R)$ (since $R = n_R/NR\Delta z$ and $\delta r = \delta n_r\Delta z$). For each layer line for which n has been determined, δn_r is calculated. Now it is unlikely that δn_r will be a constant, because it is probable that the axis of the scan was inclined to the helical axis rather than just laterally displaced by a constant distance δr. For a displacement and a rotation the value of δn_r should be consistent with a single angle of rotation $\delta\phi$ and lateral displacement δr; if possible a rotational misalignment should be avoided when the image is scanned. Layer lines which have not previously been assigned definite values for n may now be indexed with the aid of the corrected phase values. For example, if the phases of a pair of off-meridional spots are the same, n must be even and this should be used with the information that n values on layer lines should all be multiples of some basic integer, for example, $6n$ for T4 phage tail.

The transform of a helical particle may be symmetrised so that the diffraction maxima satisfy Eq. (3.45) and it is these layer lines that are extracted for image reconstruction; only information that is consistent with the chosen helical parameters is retained (Amos and Klug 1975; Smith et al. 1976).

3.6.4   The separation of the helical diffraction pattern into two components

As mentioned at the beginning of this section, a helix can be considered as a superposition of an upper and a lower surface (Klug and DeRosier 1966). Often this concept will not lead to the additional information obtained with genuine superimposed lattices. Figure 3.29a shows the diffraction pattern of a helix with exactly 5 subunits in an axial repeat ($P = 5p$). Lattice lines are now drawn through meridional diffraction spots to give effectively two

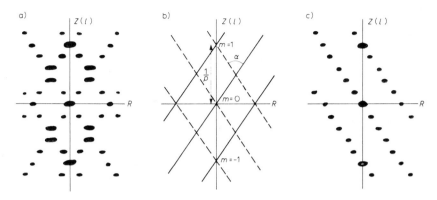

Fig. 3.29. Diffraction by helices: the separation of the top and bottom of a helix. (a) the diffraction pattern of a helix with 5 subunits/helix turn. (b) the lattice drawn through meridional points (full line ≡ top, dashed line ≡ bottom). (c) the 'one-sided' diffraction pattern. Generally complete separation is not possible because of the size of the diffraction spots and the overlap of different diffraction orders. Z (or integer l) and R define the axial and radial coordinates of the diffraction pattern and the integer m denotes the branch of the helical diffraction pattern. p is the separation between successive subunits on the helix, and α is the effective angle between the top and bottom of the helix.

lattices which are related to each other by a mirror plane along the meridian of the diffraction pattern (Fig. 3.29b). One lattice corresponds to the top of the helix (full lines) and the other to the bottom of the helix (dashed lines). These parallel lines through meridional spots as in Fig. 3.29b intersect diffraction spots with different values of n on the same branch and the procedure is equivalent to locating the branches of the diffraction pattern. If only those diffraction spots that intersect the dashed lines are included the diffraction pattern shown in Fig. 3.29c is obtained. The separation into two sets of diffraction spots is not complete because diffraction orders from different branches overlap; for example, the layer lines $l = 2$ and $l = 3$ in Fig. 3.29a include contributions from both the $m = 0$ and $m = +1$ branches (with angular harmonics J_2 and J_3). The lattice in Fig. 3.29c now looks like the reciprocal lattice for a two-dimensional crystal, where the helical parameters (l, n) are equivalent to the integers (h, k), with l along the Z-axis and n along the R-axis. The final check that the two lattices have been drawn correctly is made by tracing the set of diffraction spots from one lattice onto a transparent sheet of paper. The sheet is then turned over, maintaining the Z-axis parallel to the meridian; all the remaining spots in the diffraction pattern should be accounted for in this mirror image.

This type of separation of the top and bottom of a helix is generally not recommended for negatively-stained helical structures such as TMV, nucleoproteins and T4 phage tails, since these are essentially three-dimensional structures of excluded stain and appear as solid objects viewed from the side. It is more useful to proceed with a full helical reconstruction, removing the non-helical components of the transform. However, when a helical structure comprises a thin walled and well preserved cylinder, such as an intact microtubule, the front and back surfaces are well separated except at the curved edges. An effective separation can then be made into two lattices, corresponding to the top and bottom of the structure. Reconstructing a 'one-sided' image is then similar to separating two superimposed lattices (§ 3.7 and § 4.3).

3.7   Indexing diffraction patterns

Indexation is the determination of the simplest lattice (or set of reciprocal lattices) which accounts for *all* the diffraction maxima. A successful indexation will not only give the dimensions of the lattice but some indication of the symmetry of the molecular packing in the crystal. The reciprocal lattice is geometrically similar to the real lattice except that it is rotated by 90° and the distances measured are inverse distances.

Certain ambiguities can arise in the indexation of an optical diffraction pattern. As a result of the loss of phase information, it may be impossible to distinguish between 4- and 2-fold symmetry (P4 and P2), or 6- and 3-fold symmetry (P6 and P3). *Systematic absences* in the diffraction pattern will give further information on the crystal symmetry, such as the presence of a screw axis. It is important to note that some diffraction orders may be of low intensity, or even absent, if the Fourier transform of the motif, the basic unit of the lattice, is small or zero at a particular reciprocal lattice value. It will be remembered that the Fourier transform of a crystalline specimen is the product of the transform of the motif and the reciprocal lattice function (Eq. (3.10)). For example, the first order diffraction maxima for polyheads are either missing or of low intensity (Steven et al. 1976a and b) and it is quite easy to index the lattice incorrectly using the visible diffraction spots. It is only by noticing that some diffraction spots, particularly high order maxima, do not fit on the lattice, that it will be realised that an incorrect indexation has been made. Low resolution spots are generally not sensitive to indexation errors; but the high resolution diffraction maxima are. Other ambiguities in indexation will arise if it cannot be decided whether the image is of a simple lattice or arises from a moiré pattern resulting from the superposition of rotationally misaligned (but otherwise identical) layers. Low order spots from these superpositions often overlap and it is only high order diffraction spots that are clearly separated. Two similar lattices rotated by α can then be drawn. This separation of the two (or more) layers depends on the *linearity property* of the Fourier transform; that is, the Fourier transform of two superimposed layers $[u(r) + v(r)]$ is equal to the sum of the transforms of each layer:

$$F[u(r) + v(r)] = F[u(r)] + F[v(r)] \tag{3.49}$$

Often the two lattices u and v do not give equal intensity diffraction patterns due to, say, non-equal staining of the two layers. Image analysis will often indicate that the image does not consist of a simple lattice; so the direct interpretation of images arising from such superimposed lattices must be made with caution. It is preferable to proceed to the next stage of image processing, where the contributions from the two lattices are separated by excluding one set of diffraction maxima (§ 4.3). Applying Eq. (3.49) to the separation of superimposed layers assumes that the incident electron beam is only scattered in one of the layers. If the specimen is thick, the electron beam may be scattered in both layers and then the transform of the image cannot be indexed as two lattices. This is particularly relevant for negatively-stained specimens.

3.7.1 Simple lattices

The first step in indexation is to project the optical diffraction pattern onto graph paper and then to mark all the diffraction spots, or to make a photographic enlargement. Two sets of parallel lines are drawn, each set equidistantly spaced, so as to intersect as many diffraction spots as possible. In a simple lattice there is usually no problem in accounting for most of the diffraction maxima. Figure 3.30 shows (a) the image of a negatively-stained neuraminidase crystal of the 'square' type together with (c) its optical diffraction pattern and (d) its numerical transform amplitudes. Figure 3.31 shows the reciprocal lattice that has been drawn through the most intense spots; such an indexation is consistent with a tetragonal lattice with $a^* = b^* = 1/9.8 \text{ nm}^{-1}$, and $\alpha = 89°$, which is not too different from the ideal value of $90°$. Even in this example there are some weak diffraction maxima that cannot be accounted for in this particular construction. By rotating the diffraction pattern shown in Fig. 3.31 by $45°$ the alternative

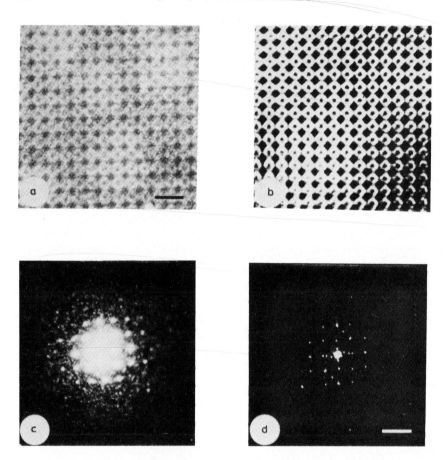

Fig. 3.30. (a) Image of a negatively-stained (1% uranyl acetate, UA) neuraminidase crystal (square lattice), (b) computer processed result (not symmetrised) using a 5 × 5 mask, (c) optical transform, (d) computer transform. 'Protein' is white. Image bar = 20 nm; diffraction bar = 0.2 nm^{-1}. (E.B. Brown, unpublished.)

indexation shown in Fig. 3.32 can be obtained, with $a^* = b^* = 1/13.5$ nm^{-1}. Although this construction does seem to account for more spots than Fig. 3.31, certain intense maxima do not fit. Figure 3.31 is probably the correct indexation because the most evident repeat distance in the micrograph is 9.8 nm rather than 13.5 nm ($\simeq \sqrt{2} \times 9.8$ nm). It is likely that the image in Fig. 3.30a contains contributions from either a second lattice or results from lattice distortions (e.g. a bent crystal). In fact the 'additional' diffraction spots are just those observed in the image transform of a neuraminidase crystal

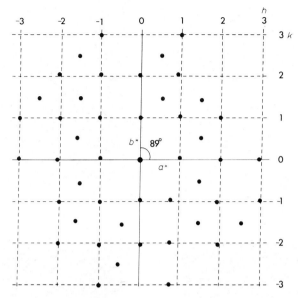

Fig. 3.31. Reciprocal lattice construction for neuraminidase 'square' crystal shown in Fig. 3.30.
$a^* = b^*$, $a = b = 9.8$ nm.

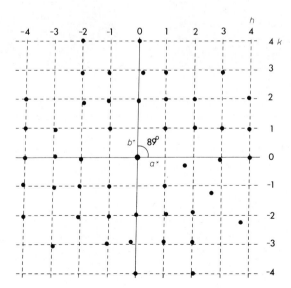

Fig. 3.32. Incorrect reciprocal lattice construction for neuraminidase 'square' crystal shown in Fig. 3.30. $a^* = b^*$, $a = b = 13.5$ nm.

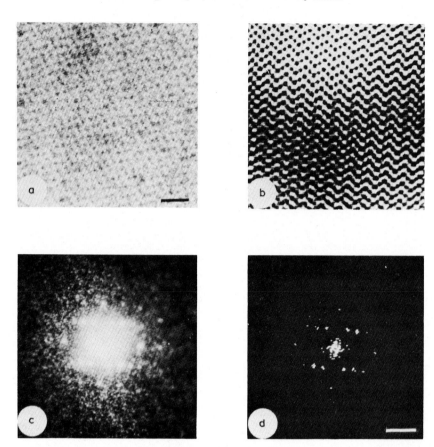

Fig. 3.33.   (a) Image of a negatively-stained (1% uranyl acetate, UA) neuraminidase crystal (zig-zag type), (b) computer processed result (no symmetrisation), (c) optical transform, (d) computer transform. Image bar = 20 nm, diffraction bar = 0.2 nm^{-1}. (E.B. Brown, unpublished.)

with a 'zig-zag' pattern (Fig. 3.33) which can be indexed as shown in Fig. 3.34; only a few random spots cannot be accounted for. The reciprocal lattice constants a^* and b^* are, respectively, $1/13.9$ nm^{-1} and $1/6.8$ nm^{-1}. It is likely that the two lattices shown in Fig. 3.31 and Fig. 3.34 do belong to the same crystal. The optical diffraction pattern of the 'zig-zag' lattice shows a dominant 4th order in the a^* direction (periodicity = 13.9 nm) and the absence of spots except at multiples of 4 indicates a 4-fold *screw-axis* in this direction*. The repeat along the other axis of the lattice is 6.8 nm which is

* Screw symmetry corresponds to a rotation and vertical translation of a subunit in the unit cell parallel to the axis of rotation; thus 4_1 symmetry indicates that the rotation of a subunit

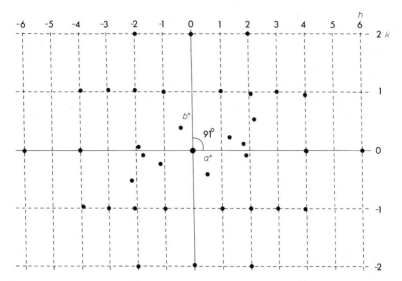

Fig. 3.34.   Reciprocal lattice construction for neuraminidase 'zig-zag' crystal shown in Fig. 3.33.
$a^* \neq b^*$, $a = 13.9$ nm, $b = 6.8$ nm.

close to the size of the square lattice multiplied by $\sqrt{2}/2$; that is, the 'zig-zag' pattern is a 45° ($\sin 45° = \sqrt{2}/2$) projection of the square lattice. The simplest crystallographic unit cell is $P4_1$ with $a = b = 9.8$ nm, $c = 13.9$ nm, but earlier analysis of the square pattern (§ 3.4.3) indicates a $P4_122$ *space group*. Confirmation of this molecular arrangement requires tilted images of the specimen.

Diffraction patterns with hexagonal symmetry are commonly found in membrane systems. Figure 3.35 shows (a) the image of a negatively-stained gap junction together with (b) its numerical transform photographed from a television display. A hexagonal lattice with $a^* = b^*$ and $\alpha = 61°$ ($\simeq 60°$) may be constructed as shown in Fig. 3.36. Certain high spatial frequency spots do not fit on the lattice, nor do they show any evidence of hexagonal symmetry, so it is unlikely that they arise from the gap junction structure. The diffraction spots marked on the lattice satisfy the relation $v = a^*\sqrt{(h^2 + hk + k^2)}$, so that a graph of v (determined from the diffractometer constant or sampling in the numerical transform) against $\sqrt{(h^2 + hk + k^2)}$

by 90° (360°/4) with a vertical translation of 1/4 of a unit cell will match the position and orientation of a second subunit in the unit cell. Generally for n_t symmetry, a rotation of 360°/n and a vertical translation by a fraction t/n of a unit cell will correspond to the position of a second subunit (Sherwood 1976).

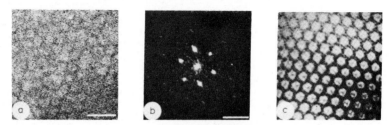

Fig. 3.35.   (a) Image of negatively-stained (1 % sodium silicotungstate, SST) gap junction from liver cells, (b) the computed transform, (c) the filtered result (no symmetrisation). Image bar = 20 nm, diffraction bar = 0.2 nm^{-1}. (N.G. Wrigley, unpublished.)

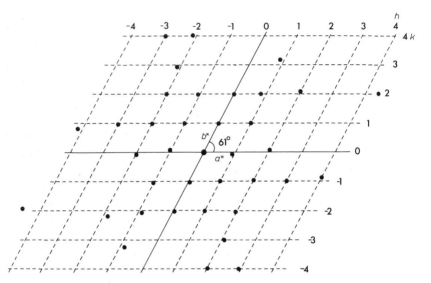

Fig. 3.36.   Reciprocal lattice construction for the gap junction shown in Fig. 3.35. $a^* = b^*$, $a = b$, $a = 1/a^*\sqrt{(4/3)} = 8.2$ nm.

will give an average value for the separation $a = \sqrt{(4/3)} \times (1/a^*)$ of the subunits making up the hexagonal lattice. In this example, all diffraction orders, including the first order, are observed, so that this type of hexagonal lattice is easy to index. However, for hexagonal lattices where certain diffraction orders are missing, the easiest way to index the diffraction pattern is in terms of radial orders; this means drawing circles about the centre of the transform, passing through sets of six diffraction spots. Figure 3.37 shows these radial orders drawn on the normal reciprocal lattice. Indexation then corresponds to attempting to fit these circles to sets of six diffraction spots,

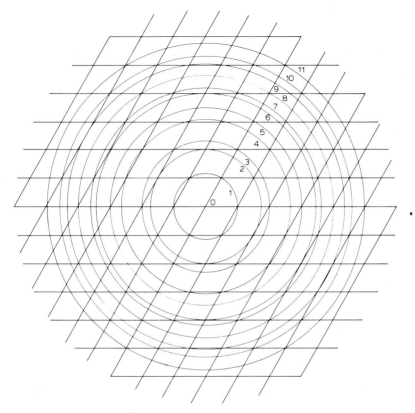

Fig. 3.37. Diagram showing the radial orders in a hexagonal lattice. Relative radii are $1 : \sqrt{3} : \sqrt{4} : \sqrt{7} : \sqrt{9} : \sqrt{12} : \sqrt{13} : \sqrt{16} : \sqrt{19} : \sqrt{21} : \sqrt{25}$. (U. Aebi, unpublished.)

for example, by projecting a transparency of the circles onto a hexagonal diffraction pattern, and adjusting the magnification of the projection to achieve the best match of circles to diffraction spots. Alternatively, circles through sets of six spots can be drawn on the diffraction pattern. A correct indexation in terms of radial orders means that the radii (or diameters) of the circles should be in the ratios:

$$1 : \sqrt{3} : \sqrt{4} : \sqrt{7} : \sqrt{9} : \sqrt{12} : \sqrt{13} : \sqrt{16} : \sqrt{19} : \sqrt{21} : \sqrt{25} \quad (3.50)$$

If certain radial orders are missing because the transform $G(v)$ of the motif is zero or small at the corresponding v values, then the ratios of successive radii will not fit the pattern shown in Eq. (3.50). Then Table 3.4, which shows the ratios of two radial orders (II/I), may be used to resolve any ambiguities

TABLE 3.4

Ratios of the radii of pairs of radial orders (II:I) for a hexagonal lattice

II \ I	1	2	3	4	5	6	7
2	1.73						
3	2.00	1.16					
4	2.65	1.53	1.33				
5	3.00	1.73	1.50	1.13			
6	3.46	2.00	1.73	1.31	1.16		
7	3.61	2.08	1.81	1.36	1.20	1.04	
8	4.00	2.31	2.00	1.51	1.33	1.16	1.11

in indexing. For example, if the first two circles drawn have a ratio 1.70 ± 0.05, then the appropriate radial orders could be 2/1, 5/2 or 6/3. The measurement of further radial ratios should then reduce the possibilities for indexation; a ratio of third to first radii of 2.00 ± 0.05 indicates radial orders 3/1, 6/2 or 8/3 but only 3/1 is consistent with the first measurement. Once all the radii have been satisfactorily fitted using Table 3.4, the lattice can be indexed in terms of (h, k). If only two or three radial orders are evident in the diffraction pattern an unambiguous indexation may not be possible; it is the high resolution diffraction orders that will be most useful for indexation. Distinguishing between a P3 and a P6 lattice is not possible from the diffraction amplitudes alone; the phase information should assist: in a P3 lattice 3-fold related diffraction peaks (120° rotation in the transform) should have the same phase, whereas for a genuine hexagonal arrangement of subunits 6-fold related transform phases (60° rotation) should be similar (International Tables for X-ray Crystallography, Vol. I).

It is quite common to find that sets of six diffraction spots lie on an ellipse rather than a circle. If the ellipse is highly eccentric then the images should be discarded until improved specimen preparation techniques produce less distortion. For small distortions, the lattice constant a can be obtained by averaging the three lattice constants obtained by measuring pairs of diametrically opposite diffraction spots.

3.7.2   Superimposed lattices

Superimposed lattices arise when an intact vesicle, cylinder or membrane dries down on a specimen grid. If the two lattices are rotated with respect to each other, it should be evident from the optical diffraction pattern, either by a blurring or by a 'doubling' of the diffraction spots. If the two lattices are merely translated and not rotated with respect to each other it will not be evident from the optical diffraction pattern; this is a result of the translational invariance of the Fourier transform amplitudes.

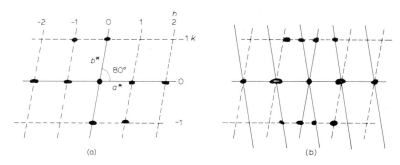

Fig. 3.38.   Reciprocal lattice for (a) opened tubes, (b) closed tubes of microtubules, showing the separation of transforms from a 'two-sided' image. (From Erickson 1974.)

Figure 3.38a shows the diffraction pattern of an open sheet (single layer) of a microtubule; the intact microtubule comprises two similar lattices with a relative rotation of 20°; Fig. 3.38b shows the diffraction pattern that results from two such layers, i.e. the closed tube (Erickson 1974). Images like this are sometimes referred to as 'two-sided' images. Separating and indexing the two lattices involves determining which diffraction maxima belong to which lattice. For example, Fig. 3.38b is quite difficult to index because the information is confined to relatively low diffraction orders. The separation of the two lattices and production of an image of a single lattice is referred to as 'one-sided' filtering. For example, the optical diffraction pattern in Fig. 3.39a of the image of a negatively-stained T4 polyhead (Fig. 3.39b) shows clear evidence of two superimposed lattices, misaligned by about 10°. This diffraction pattern is very difficult to index because the 1st and 3rd radial orders are absent (Steven et al. 1976a). If indexation is based on the assumption that the first sextet of diffraction spots at $1/5.6$ nm^{-1} corresponds to the first radial order, it is possible to account for most diffraction peaks using two rotated lattices. However, some of the high resolution spots fail to fit on such a lattice. The use of Table 3.4 helps to narrow down the possible indexations: there are three radial orders shown in Fig. 3.40a with ratios $2:1 = 1.60$ and $3:1 = 1.95$. The radial orders that best fit these ratios are 2, 4 and 6. Based on this, the diffraction pattern of a

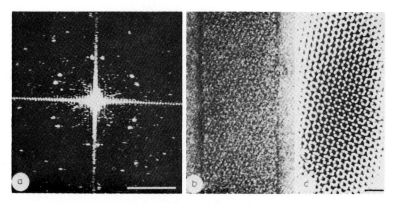

Fig. 3.39. (a) Optical transform, (b) micrograph, (c) 'one-sided' optical reconstruction of a negatively-stained (2%, sodium phosphotungstate, SPT) coarse T4 polyhead. Image bar = 20 nm, diffraction bar = 0.2 nm^{-1}. (From Steven et al. 1976a.)

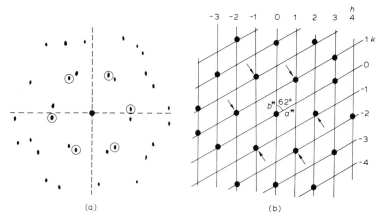

(a)   (b)

Fig. 3.40. Reciprocal lattice for the coarse polyhead in Fig. 3.39; the first order of the hexagonal lattice is missing. (a) original optical diffraction pattern with one-sided spots ringed. (b) reciprocal lattice drawn through the spots resulting from one-sided diffraction (arrowed). (h, k) define the diffraction order; a^* and b^* are the reciprocal lattice constants. For a hexagonal lattice $a^* = b^*$. (From Steven et al. 1976a.)

single lattice is shown in Fig. 3.40b. In a study of different mutants of T4 polyheads, Steven et al. (1976a and b) attempted to detect by electron microscopy, structural differences in the arrangement of the proteins on the surface of the polyheads. Before this could be achieved a set of images, together with their optical diffraction patterns (Fig. 3.41) had to be analysed

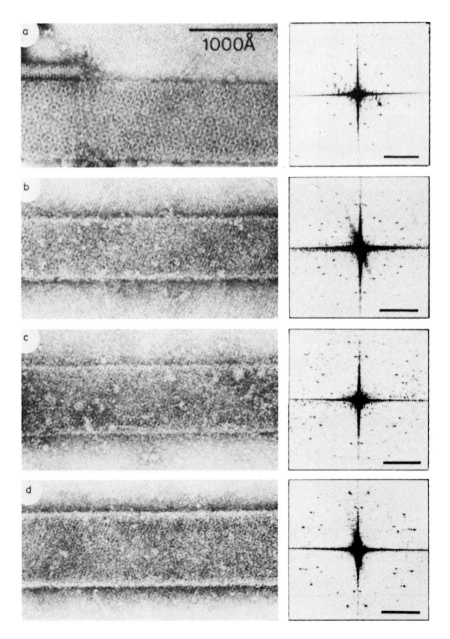

Fig. 3.41. Electron micrographs and optical diffraction patterns of four different types of T-even polyheads. (a) coarse polyhead, (b) A-type, (c) B-type, (d) C-type. Image bar = 100 nm, diffraction bars = $0.2\,\text{nm}^{-1}$. (From Steven et al. 1976b.)

and the diffraction patterns indexed correctly, before one-sided images could be produced for structural interpretation. Unfortunately the indexation schemes for different types of polyhead are not the same; the dominant orders for the coarse type (a) are 2 and 4; for type A (b) 2,3 and 4; for type B (c) 2, 3 and 4, and for type C (d) 4 and 7. The final indexation is shown in Fig. 3.42,

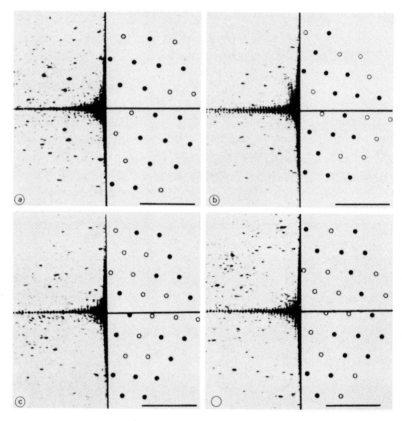

Fig. 3.42.   Indexation figure for the different polyheads: (a) coarse, (b) A, (c) B, (d) C. In each case half of the diffraction pattern is shown and the other half shows the indexation of one of the two lattices. Solid circles are visible diffraction orders, open circles are points where there is no significant intensity. The second lattice can be generated by reflection in the equator. Diffraction bars = $0.2 \, nm^{-1}$. (A.C. Steven, unpublished.)

where the left half of each figure shows the original diffraction pattern from two superimposed lattices and the right half shows that from a single lattice. The diffraction maxima observed are denoted by solid circles and open circles are positions in the reciprocal lattice where there is no significant intensity. These images and diffraction patterns form the basis of a particularly good example of the study of the arrangement and assembly of these mutants, which will be complementary to a biochemical investigation of the morphogenesis of the T4 phage head.

3.7.3 Helical particles

In indexing a helical diffraction pattern first find the meridional diffraction maxima (corresponding to $n = 0$); the altitudes of the meridionals then give the axial repeat p. The problem is now to locate the layer lines between the equator and the meridionals; all diffraction maxima must be consistent with a single layer line spacing. Remember that the helix is essentially a one-dimensional lattice and, although the diffraction pattern is discrete in the Z-direction, it is continuous in the R-direction with layer line intensities proportional to $[J_n(2\pi R r_0)]^2$ (§ 3.6.1). Select the most evident layer lines; for the T4 phage tail, this corresponds to layer lines $l = 2, 3, 4$ and 5 (e.g. Fig. 3.43a; Smith et al. 1976). Their altitudes Z are then measured. Diffraction

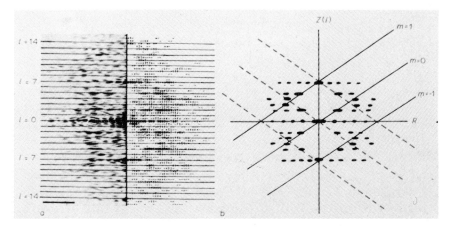

Fig. 3.43. (a) Optical/computer transform of a T4 phage tail (uncontracted) showing the layer line indexation. (b) selecting only those layer lines with a clear structure (2, 3, 4, 5 and 7), the lattice is indexed to be consistent with the selection rule $l = -2n + 7m$ where only n values which are a multiple of 6 are allowed. Diffraction bar $= 0.2$ nm^{-1}. (From Smith et al. 1976.)

maxima on different layer lines with the same value of R should correspond to angular orders J_n and J_{-n}, such as $l = 2$ and 5 in Fig. 3.43a. As a check the sum of the altitudes of layer lines corresponding to J_n and J_{-n} should be equal to the altitude of the first meridional. If there is no meridional then the only way to determine p is to look for two off-meridional lines that are clearly J_n and J_{-n}, although the values of n may not yet be known. In order to obtain information on the values of n, it is a good idea to look at the original micrograph; for example, an end-on view of the T4 tail or a T4 fragment will show the 6-fold rotational symmetry of the annuli comprising the helix, and it is probable that only angular orders J_{6n} can contribute to the layer lines. It should now be possible to draw all the layer lines on the diffraction pattern: the layer lines must all be equidistant, and their altitudes must be a multiple of some constant $(1/c)$. The values of n on each layer line must all be a multiple of some basic value; for the T4 phage tail the angular orders must be a multiple of 6, that is, 0, 6, 12, 18, ... This helps to restrict the allowed values of n. On each layer line the R value of the first diffraction peak R_{max} corresponds to the first maximum of $J_n(2\pi R r_0)$. Since $J_n(X)$ are tabulated, an estimate of $2\pi R_{max} r_0$ will enable an estimate of n to be made. In Fig. 3.43a, the near-meridional maxima on layer lines $l = 2, 5$ correspond to an R_{max} value of $0.12\ \mathrm{nm}^{-1}$. The diameter of the particle as estimated from the electron micrograph $(2r_0)$ is approximately 19 nm, and the value of $2\pi R_{max} r_0 = 7.2$. From Table 3.5 this indicates a value for n between 5 and 6.

TABLE 3.5

Position of the first maximum of $J_n(X)$ for $n = 0\text{--}12$

n	0	1	2	3	4	5	6	7	8	9	10	11	12
X	0.0	1.8	3.1	4.2	5.3	6.4	7.5	8.6	9.7	10.7	11.8	12.8	13.9

The next pair of off-meridionals that are measurable correspond to $l = 3$ and 4 with $R_{max} = 0.24\ \mathrm{nm}^{-1}$; thus $2\pi R r_0 = 14.4$ which indicates a Bessel function of order 12 or 13. These two calculations are consistent with angular harmonics of order $6n$, J_{6n}; if J_6 contributes to $l = 2$ then $l = 5$ corresponds to J_{-6}, whilst $l = 3, 4$ correspond to $J_{\pm 12}$. The layer lines $l = 1, 6$ cannot be measured but it is most likely that the angular orders on these layer lines are $J_{\pm 18}$, which would explain their low intensity. Note that the meridionals always correspond to J_0 and they do not give any information on possible n values; the altitudes of the meridionals give one basic parameter of the helix, p. The success in estimating n values depends on the uniform staining of the three-dimensional particle so that the r_0 in $J_n(2\pi R r_0)$ corresponds to the radius of the particle measured on the micrograph. Flattened or distorted helices do not have a single radius. Moody (1967) gives a table of the first diffraction maxima for a flattened helix, a circular helix (as in Table 3.5), a one-sided helix (stained only on the top or bottom) and a flattened one-sided helix; one-sided helices are recognisable by the asymmetry of the diffraction pattern along the R-axis.

A more accurate determination of the helical parameters and a complete indexation is obtained by fitting the layer line altitudes Z by the equation (Smith et al. 1976):

$$Z = \frac{1}{p}\left(\frac{\Omega n}{2\pi} + m\right) \tag{3.51}$$

when m is the integer defining a particular branch (cross) of the helical diffraction pattern. Thus in Fig. 3.43b, the $m = 0$ branch passes through the origin of the diffraction pattern whilst the $m = \pm 1$ branches pass through the meridionals corresponding to $Z = \pm 1/p$. For $m = 0$ several values of the screw angle of the helix Ω can be obtained from Eq. (3.51) using $\Omega = (Z/1/p)(360°/n)$; other branches $m = \pm 1$ will give other values of Ω and an average can be calculated. Alternatively a least squares procedure can be used to estimate Ω by minimising the residual:

$$\Delta = \sum_l (Z_l - an_l - bm_l)^2 W_l \tag{3.52}$$

where Z_l is the measured layer line altitude, n_l is the order of the angular harmonic sampled on the l th layer line and m_l is the label of the branch; W_l is a weighting factor reflecting the relative accuracy of the altitude measurement. a ($\Omega/2\pi p$) and b ($1/p$) are chosen to minimise the residual Δ. Ω is then determined from $2\pi a/b$, whilst the value of the axial repeat distance p is determined from $1/b$ (Smith et al. 1976).

An alternative, sometimes complementary, way of indexing a helical diffraction pattern is to produce an (n, l) plot which is similar to the (h, k) plot for a two-dimensional lattice. Figure 3.44a shows an (n, l) plot for a basic helix with 7 units in two complete turns ($c = 7p$). Such a plot assists in the assignment of the branch values m; the $m = 0$ branch always passes through the origin $(0, 0)$ whilst different branch lines intersecting sets of diffraction maxima must all be parallel as shown in Fig. 3.44a. Drawing this type of diagram closely corresponds to finding a reciprocal lattice for a two-dimensional crystal. Figure 3.44b shows the corresponding (n, l) plot for a helical particle comprising six such basic helices; the layer line spacing

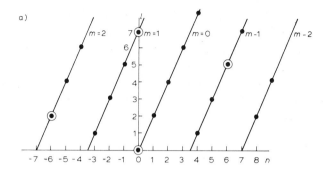

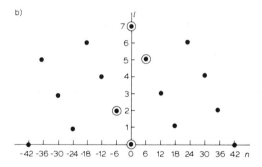

Fig. 3.44.   Diffraction by helices: (a) (n, l) plot for a basic helix with 7 units in two turns, $c = 2P$, $c = 7p$; selection rule $l = -2n + 7m$, (b) six such helices related by a 6-fold axis. The spacing of the layer lines is unaltered, but Bessel functions whose order are a multiple of six appear (J_0, J_6, J_{12} etc.). Ringed spots correspond to those that appear in the original (a). Note the change in the n scale. (R.A. Crowther, unpublished.)

is unaltered (c is unchanged, $Z = l/c$) but only Bessel functions whose order is a multiple of six appear.

References

Abramowitz, M. and I.A. Stegun (1965), Handbook of mathematical functions (Dover Publications, New York), Tables 9.1–9.3.

Aebi, U., P.R. Smith, J. Dubochet, C. Henry and E. Kellenberger (1973), A study of the structure of the T-layer of *Bacillus brevis*, J. Supramol. Struct. *1*, 498–522.

Agar, A.W., R.H. Alderson and D. Chescoe (1974), Principles and practice of electron microscope operation, in: Practical methods in electron microscopy, Vol. 2, A.M. Glauert ed. (North-Holland, Amsterdam).

Amos, L.A. and A. Klug (1975), Three-dimensional image reconstruction of the contractile tail of T4 bacteriophage, J. Mol. Biol. *99*, 51–73.

Beeston, B.E.P. (1972), An introduction to electron diffraction, in: Practical methods in electron microscopy, Vol. 1, part 2, A.M. Glauert, ed. (North-Holland, Amsterdam), pp. 193–323.

Beorchia, A. and P. Bonhomme (1974), Experimental studies of some dampings of electron microscope phase contrast transfer functions, Optik *39*, 437–442.

Cochran, W., F.H.C. Crick and V. Vand (1952), The structure of synthetic polypeptides. I. The transform of atoms on a helix, Acta Crystallogr. *5*, 581–586.

Cowley, J.M. and A.F. Moodie (1957), The scattering of electrons by atoms and crystals. I. A new theoretical approach, Acta Crystallogr. *10*, 609–619.

Crowther, R.A. and L.A. Amos (1971), Harmonic analysis of electron microscope images with rotational symmetry, J. Mol. Biol. *60*, 123–130.

Crowther, R.A. and U.B. Sleytr (1977), An analysis of the fine structure of the surface layers from two strains of *Clostridia*, including correction for distorted images, J. Ultrastruct. Res. *58*, 41–49.

DeRosier, D.J. and A. Klug (1968), Reconstruction of three dimensional structures from electron micrographs, Nature *217*, 130–134.

DeRosier, D.J. and P.B. Moore (1970), Reconstruction of three-dimensional images from electron micrographs of structures with helical symmetry, J. Mol. Biol. *52*, 355–369.

Erickson, H.P. (1974), Microtubule surface lattice and subunit structure and observations on reassembly, J. Cell Biol. *60*, 153–167.

Frank, J. (1973), Computer processing of electron micrographs, in: Advanced techniques in biological electron microscopy, J.K. Koehler, ed. (Springer-Verlag, Berlin), pp. 215–274.

Grinton, G.R. and J.M. Cowley (1971), Phase and amplitude contrast in electron micrographs of biological material, Optik *34*, 221–233.

Harburn, G., C.A. Taylor and T.R. Welberry (1975), Atlas of optical transforms (Bell and Hyman Ltd., London).

Henderson, R. and P.N.T. Unwin (1975), Three-dimensional model of purple membrane obtained by electron microscopy, Nature *257*, 28–32.

Horne, R.W. and R. Markham (1972), Application of optical diffraction and image reconstruction techniques to electron micrographs, in: Practical methods in electron microscopy, Vol. 1, part 2, A.M. Glauert, ed. (North-Holland, Amsterdam), pp. 327–434.

International tables for X-ray crystallography (1969), N.F.M. Henry and K. Lonsdale, eds. (The Kynoch Press, Birmingham), Vol. I, Symmetry Groups.

Klug, A. and D.J. DeRosier (1966), Optical filtering of electron micrographs: reconstruction of one-sided images, Nature *212*, 29–32.

Klug, A., F.H.C. Crick and H.W. Wyckoff (1958), Diffraction by helical structures, Acta Crystallogr. *11*, 199–213.

Lenz, F.A. (1971), Transfer of image information in the electron microscope, in: Electron microscopy in material science, U. Valdrè ed. (Academic Press, New York), pp. 541–569.

Moody, M.F. (1967), Structure of the sheath of bacteriophage T4. I. Structure of the contracted sheath and polysheath, J. Mol. Biol. *25*, 167–200.

Mulvey, T. (1973), Instrumental aspects of image analysis in the electron microscope, J. Microsc. (Oxford) *98*, 232–250.

Pereira, H.G. and N.G. Wrigley (1974), *In vitro* reconstitution, hexon bonding and handedness of incomplete adenovirus capsid, J. Mol. Biol. *85*, 617–631.

Ross, M.J., M.W. Klymkowsky, D.A. Agard and R.M. Stroud (1977), Structural studies of a membrane-bound acetylcholine receptor from *Torpedo californica*, J. Mol. Biol. *116*, 635–659.

Sherwood, D. (1976), Crystals, X-rays and proteins (Longman Group, London).

Smith, P.R., U. Aebi, R. Josephs and M. Kessel (1976), Studies of the structure of the T4 bacteriophage tail sheath – I. The recovery of three-dimensional structural information from the extended sheath, J. Mol. Biol. *106*, 243–275.

Steven, A.C., U. Aebi and M.K. Showe (1976a), Folding and capsomere morphology of the P23 surface shell of bacteriophage T4 polyheads from mutants in five different head genes, J. Mol. Biol. *102*, 373–407.

Steven, A.C., E. Couture, U. Aebi and M.K. Showe (1976b), Structure of T4 polyheads – II. A pathway of polyhead transformations as a model for T4 capsid maturation, J. Mol. Biol. *106*, 187–221.

Unwin, P.N.T. and R. Henderson (1975), Molecular structure determination by electron microscopy of unstained crystalline specimens, J. Mol. Biol. *94*, 425–440.

Unwin, P.N.T. and C. Taddei (1977), Packing of ribosomes in crystals from the lizard, *Lacerta sicula*, J. Mol. Biol. *114*, 491–506.

Wrigley, N.G. (1968), The lattice spacing of crystalline catalase as an internal standard of length in electron microscopy, J. Ultrastruct. Res. *24*, 454–464.

Wrigley, N.G., J.J. Skehel, P.A. Charlwood and C.M. Brand (1973), The size and shape of influenza virus neuraminidase, Virology *51*, 525–529.

Chapter 4

Image processing

The step following image analysis is image processing, in which the information consistent with the specimen structure, as determined by analysis, is used to produce an enhanced image. For example, for a two-dimensional crystal showing *translational symmetry*, the only information retained corresponds to the diffraction maxima and all non-periodic information is removed from the transform. A new image is then reconstructed from these selected Fourier coefficients. Clearly care has to be taken that the information used is correct, otherwise a reconstructed image that is visually satisfying but structurally incorrect can be produced. This is particularly relevant when attempting to separate diffraction maxima arising from a 'two-sided' image (§ 4.3); incorrect indexation and the use of diffraction spots from both layers of the structure will lead to a reconstructed image with false detail. It is important to verify that the reconstructed image is consistent both with its original and with any other information such as biochemical data obtainable about the specimen.

   Most of the image processing methods described in this chapter are based on the *spatial filtering* of the image transform; that is, the selection of certain spatial frequencies from the transform to calculate a reconstructed image. However, any spatial filtering technique has its real space equivalent (Aebi et al. 1973). For example, in the transform of a two-dimensional periodic structure, a reconstructed image is calculated using only those parts of the transform that are consistent with the periodicity; all other Fourier coefficients are excluded. The real space equivalent is an average obtained by superimposing nominally identical subunits photographically, using the original micrograph together with a suitable translation of the image on photographic paper (Horne and Markham 1972; see § 4.2). Similarly *rotational filtering* may be achieved by selective Fourier filtering, using only those Fourier coefficients that show n-fold rotational symmetry to produce the image (Crowther and Amos 1971). With the original micrograph this may be done photographically by superimposing an image of the particle n times, rotating the image about the centre of the particle for each printing by $360°/n$ (Horne and Markham 1972; see § 4.1.1). In general, spatial filtering of an image transform is equivalent to some type of averaging procedure using the original micrograph. If facilities, particularly computational, are limited then real space techniques should be considered, based on information obtained by image analysis (e.g. the repeat distance of the structure). Fourier techniques, however, are more flexible and for some procedures, such as rotational filtering, far more objective. In addition, numerical transforms yield not only intensity information but also phase information (§ 3.4.3), which can lead to improvements in the reconstructed image that cannot be obtained by photographic methods alone.

The purpose of image processing is to enhance features in the original image, but this must always be done objectively. It is not much use to process an image at 1 nm resolution, when the optical transform clearly indicates that structural information ceases at 2 nm; all that will be added is fine detail that can in no way be related to the structure of the specimen. If high resolution information is not present in the first place, image processing cannot produce a structurally significant high resolution result. Its main purpose is to extract all the structural information from the image, making it more visible. In mathematical terms, the *signal-to-noise* ratio in the image is increased, where the *signal* corresponds to the structural component of the image and the *noise* is non-structural contributions, such as the carbon support film, effects of radiation damage, inelastic electron scattering and statistical noise on the photographic emulsion used to record the image. Image analysis will give an indication of which information is structural, particularly if the specimen exhibits some type of symmetry. Statistically the signal-to-noise ratio, S/N, gives a measure of the confidence with which a structure can be distinguished from its background. A S/N value of q gives a probability:

$$p = q/(1 + q) \qquad (4.1)$$

with which a particular structural detail can be distinguished from the noise. For normal microscopy a S/N value of about 5 is considered to be the minimum for a visual assessment of an image, because this gives an 83 %

probability that a structure can be recognised reliably. However, if the specimen comprises a large number of subunits arranged in a regular array containing $k \times k$ subunits the value of S/N acceptable in the image is reduced by a factor k, since by averaging over $k \times k$ subunits, the signal-to-noise ratio can be increased by a factor $\sqrt{k^2} = k$ (Unwin and Henderson 1975). Thus a statistically noisy image of a two-dimensionally ordered specimen, although visually unacceptable, may be a useful image of the specimen. A visually acceptable image of a heavily stained specimen may be a poor representation of the structure (§ 7.1).

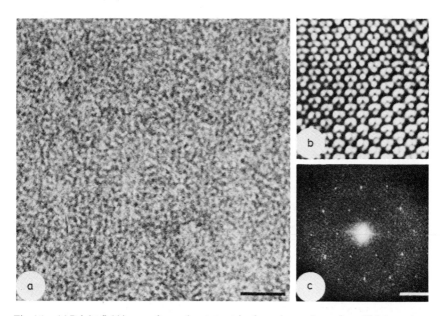

Fig. 4.1. (a) Bright-field image of uranyl acetate-stained purple membrane from *Halobacterium halobium* together with its optical transform (c). (b) shows the enhanced image, printed at the same magnification as the original (a), obtained by retaining only image information that is consistent with a hexagonal lattice (unit cell, 6.2 nm × 6.2 nm). The 'protein' is shown as white in (b). (a, b) image bar = 20 nm, (c) transform bar = 0.2 nm^{-1} and the highest order diffraction spot corresponds to 0.5 nm^{-1} or a resolution of 2 nm. (M.V. Nermut, unpublished.)

The importance of a high signal-to-noise ratio is illustrated in Fig. 4.1a which shows an image of a lightly stained single layer of purple membrane from *Halobacterium halobium* supported on a thin carbon film (about 10 nm thick). The hexagonal arrangement of the protein subunits is not evident because the signal-to-noise ratio is less than unity ($p < 50\%$ from Eq. (4.1)).

However, the optical transform (Fig. 4.1c) clearly shows a hexagonal array of diffraction spots extending to 2 nm resolution ($v = 0.5$ nm^{-1}) superimposed on a diffuse background arising from the carbon film and from non-uniformity of the stain. The 'signal' corresponds to the diffraction spots and the 'noise' is the remainder of the transform. The intense spot at the centre of the transform arises from the uniform background in the bright-field image. By using only the diffraction maxima and excluding all information not consistent with a hexagonal lattice to reconstruct the image, the result shown in Fig. 4.1b is obtained. The hexagonal arrangement of the 'protein' subunits (shown as white – the stain exclusion pattern) is clearly visible. This filtering operation was equivalent to averaging over about 6 × 6 unit cells in the original image and the signal-to-noise ratio in the reconstructed image is about 5. Averaging over a larger area of the original image is possible, but in this image the lattice was distorted from exact hexagonal symmetry, shown by the broadening of the diffraction spots in Fig. 4.1c and the variation in the appearance of the protein subunits in Fig. 4.1b (see § 4.2). Further averaging might have produced an image with a higher signal-to-noise ratio but would be no more informative about the subunit structure.

The improvement in the signal-to-noise ratio which leads to an enhancement of the original image requires symmetry in the specimen. Thus rotational averaging produces an improvement in S/N by a factor of $\sqrt{n}$ for n-fold rotational symmetry (see § 4.1).

Image processing should not be considered to be a technique for enhancing poor images; it may succeed but it is not an alternative to taking the best possible images of good specimen preparations. Practical ways of improving the S/N of the original image should also be used. For example, a thin carbon support film, careful staining of the specimen and the use of minimal radiation conditions are all factors within the control of the microscopist. It is sensible to use a carbon film that is significantly thinner than the specimen, otherwise an unnecessary noise contribution is added to the image (Johansen 1976).

For non-periodic specimens image processing has a limited scope. If the transform of an image can be clearly separated into low resolution (low spatial frequencies) and high resolution (high spatial frequencies) regions, then there is still the possibility of reducing the contribution of the noise to the image by selective spatial filtering (§ 4.7). However, the principal way of enhancing images from non-periodic specimens involves instrumental techniques such as STEM and dark-field microscopy (see Chapter 5).

4.1 Specimens with rotational symmetry

There are many examples of biological particles which when seen in a particular orientation show n-fold rotational symmetry; that is, a rotation of $360°/n$ of the particle about its centre leaves the appearance of the particle unchanged. By superimposing n images of the same particle each rotated by $360°/n$, the noise, which does not display rotational symmetry, is reduced by a factor of $\sqrt{n}$.

4.1.1 Photographic superposition

The simplest way to do rotational filtering is by photographic superposition (Horne and Markham 1972). The image is projected onto photographic paper at a convenient enlargement; e.g. so that the particle diameter is about 5–10 cm. The photographic paper is rotated about the centre of the particle, making an exposure after each rotation of $360°/n$ for $1/n$th of the exposure time required to make a normal photographic print of the particle. In order to produce an accurate rotation of the photographic paper, two cardboard discs about 20 cm in diameter are pinned together at their centres; one disc is marked with an angular scale marked in $1°$ intervals, whilst the second disc is marked with a pointer. The centres of the discs are aligned with the rotational centre of the particle image and the photographic paper is fixed to the inner disc and an exposure made. The inner disc is then rotated by $360°/n$ and n exposures made. The result of rotational filtering for a group of nine hexons from adenovirus ($n = 3$) is shown in Fig. 4.2; three successive exposures were made by rotating the photographic paper by $120°$ each time. The three-fold sub-division of each hexon is much clearer in the reconstruction (Fig. 4.2b) than in the original image (Fig. 4.2a); the increase in S/N in this example was only $\sqrt{3} = 1.73$.

The choice of n is subjective and care should be taken with photographic averaging. For the groups of nine hexons, $n = 3$ is fairly evident but sometimes the choice of n is not obvious. For example, the protein disc from tobacco mosaic virus (TMV) has a very high order of rotational symmetry ($n = 17$), and visually it is not possible to make an objective selection from the results obtained by making $n = 16$, 17 or 18 superpositions (Crowther and Amos 1971). There are certain precautions that can be taken to ensure that the correct value for n has been chosen. If only $n - 1$ superpositions are made, then the remaining $360°/n$ sector will be similar to the other $n - 1$ sectors only if the correct n has been chosen. Also if the rotational symmetry

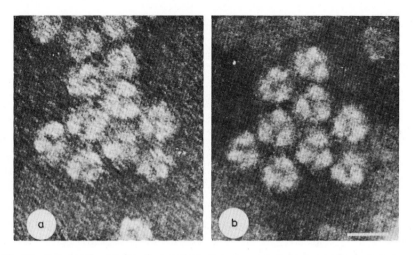

Fig. 4.2.   Rotational averaging of a negatively-stained group of nine hexons from adenovirus: (a) original, (b) 3-fold photographic superposition. Image bar = 10 nm. (J.V. Heather, unpublished.)

is of a high order and can be factorised, then the images resulting from *n*-fold superposition and its factors will give similar results; for example if $n = 12$, then $n = 3$, $n = 4$ and $n = 6$ superpositions will all produce similar structures, but $n = 8$ will produce a different result. In principle, if the value of *n* is in doubt, rotational filtering can be made for all possible *n* values and a visual assessment may enable the correct choice of *n* to be made; with the TMV protein disc $n = 16$ may be eliminated because $n = 8$ does not give a similar result. However, a more objective method for the choice of *n* may be made using numerical techniques with either the image or its transform (Crowther and Amos 1971). It is not possible to use the optical transform of the image for rotational filtering because it is very difficult to select only those Fourier components which correspond to *n*-fold symmetry as a result of the absence of phase information (§ 3.5).

4.1.2   Computer methods: analysis

Numerical rotational filtering requires the image of the particle to be transformed from Cartesian (x, y) to polar (r, ϕ) coordinates ($x = r \cos \phi$, $y = r \sin \phi$), with the origin of the polar coordinates corresponding to the rotational centre of the particle (e.g. Fig. 3.25; § 3.6). With this coordinate system the particle density exhibits periodicity in the angle ϕ; that is,

$\rho(r, \phi) = \rho(r, \phi + 2\pi)$. The density variations $\rho(r, \phi)$ can be expanded as a Fourier series in ϕ:

$$\rho(r, \phi) = \sum_{n=-\infty}^{n=+\infty} g_n(r) \exp(in\phi) \tag{4.2}$$

where n denotes the angular harmonics of the image; a particle with m-fold rotational symmetry will have strong Fourier components $g_n(r)$ corresponding to $n = 0, m, 2m, 3m, \ldots$. The Fourier coefficients at a particular radius r from the centre of the particle can be calculated using:

$$g_n(r) = \frac{1}{2\pi} \int_0^{2\pi} \rho(r, \phi) \exp(-in\phi) \, d\phi \tag{4.3}$$

Both Eqs. (4.2) and (4.3) assume that the rotational centre of the particle has been determined.

Thus rotational filtering could be implemented numerically by *interpolating* the image onto polar coordinates with the origin at the rotational centre of the particle, and calculating $g_n(r)$ at r values increasing from the centre of the particle to its outer radius. Then an image containing only particular values of n can be produced by using Eq. (4.2). However, it would be necessary to repeat this procedure several times in order to make the best choice of the rotational centre (origin of polar coordinates); a visual choice will not usually be sufficiently accurate. This means that the image must be reinterpolated for each choice of origin until the *power spectrum* (the sum of squared moduli of the $g_n(r)$) for the particular choice of m is a maximum; that is, the sum:

$$\sum_n \int_0^{r_0} |g_n(r)|^2 \, 2\pi r \, dr = P \tag{4.4}$$

taken from the centre of the particle ($r = 0$) to its outer radius ($r = r_0$) and over m-fold components only is maximised. The power of a particular harmonic n is defined as:

$$P_n = \varepsilon_n \int_0^{r_0} |g_n(r)|^2 \, 2\pi r \, dr \tag{4.5}$$

where $\varepsilon_n = 1$ for $n = 0$ and $\varepsilon_n = 2$ for $n > 0$ (to account for positive and negative values of n in the summation (4.2), since $|g_n(r)| = |g_{-n}(r)|$).

Thus it is possible to plot the rotational power spectrum P_n against n for a particular particle; for a particle with m-fold symmetry, the values of P_n corresponding to $n = m, 2m, 3m, \ldots$ should be significantly larger than the other P_n values. Such power spectra for images of groups of nine hexons are shown in Fig. 4.3 for (a) a well-preserved particle, and (b) a poorly-preserved

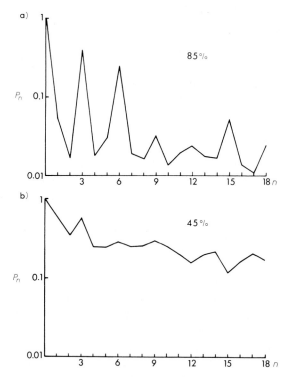

Fig. 4.3.   Rotational power spectra for groups of nine hexons from adenovirus: (a) well-preserved particle ($P = 85\%$), (b) poorly-preserved particle ($P = 45\%$).

particle. It will be seen in Fig. 4.3a that the P_n values corresponding to $n = 3, 6, 9, 12$ and 15 are significantly greater than other values of P_n; note that the P_n scale is logarithmic and P_0 is arbitrarily normalised to unity. This should be contrasted with Fig. 4.3b which shows the power spectrum of the image of a particle which is significantly distorted from 3-fold rotational symmetry. An overall assessment of m-fold rotational symmetry can be obtained by calculating the ratio of the value of P for m-fold rotational components only to the value of P calculated for all n values; this should be significantly larger than $100/m \%$, the percentage rotational power of a completely random structure (i.e., no dominant rotational components). In Fig. 4.3a the value of 85% clearly demonstrates the dominance of 3-fold rotational symmetry, whereas the value of 45% in Fig. 4.3b is not significantly greater than the 33% for a random structure.

In practice it is easier to perform rotational filtering using the Fourier transform of the image, because it avoids the necessity of interpolating the image into polar coordinates for each choice of rotational centre. Altering the origin by $(\delta x, \delta y)$ in the original image corresponds to a phase shift $\delta\omega = -2\pi(v_x\delta x + v_y\delta y)$ in the transform. By working in transform space it is necessary only to multiply all transform coefficients by $\exp(i\delta\omega)$ to effect an alteration of origin by $(\delta x, \delta y)$. This determination of an origin to maximise m-fold rotational symmetry components corresponds to the choice of the phase origin of the transform; m-fold related amplitudes and phases in the transform should be identical.

An area of image is scanned so as to include the rotational origin of the particle as near to the centre of the scan as possible; a line-printer display of the scanned image usually enables an approximate particle centre (x_0, y_0) to be chosen. The image is *floated* to a mean density of zero before an FFT is calculated (§ 3.4.2). Sometimes before these operations it is convenient to reduce noise from the background area contributing to the image by *boxing* the particle; this corresponds to extracting from the original scanned image a square area of the image that just encompasses the particle. A significant improvement is then obtained in the rotational power spectrum as a result of reducing the non-rotationally symmetric background. The image transform is transferred to a transform in polar coordinates $R, \Phi(v_x = R\cos\Phi, v_y = R\sin\Phi)$. This image transform $F(R, \Phi)$ is then expressed as a Fourier series in Φ:

$$F(R, \Phi) = \sum_{n=-\infty}^{n=+\infty} G_n(R) \exp\left[in(\Phi + \pi/2)\right] \tag{4.6}$$

where the coefficients $G_n(R)$ are related to the $g_n(r)$ coefficients by a Fourier–Bessel transform (corresponding to the Fourier transform in polar coordinates):

$$G_n(R) = \int_0^{r_0} g_n(r) J_n(2\pi rR) 2\pi r \, dr \tag{4.7}$$

and the inverse relationship is:

$$g_n(r) = \int_0^{\infty} G_n(R)\, J_n(2\pi r R)\, 2\pi R\, dR \qquad (4.8)$$

The factor $\exp(in\pi/2)$ in Eq. (4.6) is a result of the mathematical transformation to polar coordinates and has no physical significance.

The transform $F(R, \Phi)$ is multiplied by the phase factor $\exp[-2\pi i(v_x x_0 + v_y y_0)]$ transformed to polar coordinates to produce an image transform with approximately the correct phase origin. The phase origin is refined by choosing origin shifts $(\delta x, \delta y)$ about the initial origin so that the *residual*, that is, the sum of the squares of the differences (over all values of R and Φ) between m-fold related Fourier coefficients, is minimised (Fig. 4.4). This

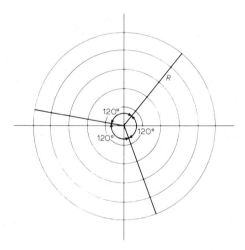

Fig. 4.4.   Rotational filtering in Fourier space for 3-fold rotational symmetry. R is the Fourier equivalent of r.

refinement maximises the m-fold symmetry in the transform. Because the amplitudes of the transform are translationally invariant, it is the symmetrisation of the phases that is the important step in rotational filtering using the image transform; thus the phase origin refinement is an essential step for successful rotational filtering. Usually the origin shifts are specified in terms of image sample points; thus the initial shift to place the origin near the centre of the particle (x_0, y_0) corresponds to a shift of (n_x, n_y) sample points and a phase shift of $-2\pi(pn_x + qn_y)/N$ for a scan size of $N \times N$ points, where p and q correspond to the coordinates (v_x and v_y) in the image transform. Subsequent origin refinements ($\delta x, \delta y$) correspond to a shift of ($\delta n_x, \delta n_y$) sample points and a phase shift of $\delta \omega = -2\pi(p\delta n_x + q\delta n_y)/N$.

There is one important decision that should be made when calculating the residual for maximum m-fold symmetry, and that is the choice of R_{max} – the maximum value of R that is to be used in origin refinement. Remembering that R corresponds to a spatial frequency in the image, large R values refer to fine structure in the original image. R_{max} should be restricted to a value that is consistent with the preservation of the structure, because using a large value for R_{max} will only add non-structural information, i.e. noise, to the reconstructed image. In negatively-stained specimens R_{max} should not be larger than 0.5 nm^{-1} (resolution 2 nm): in terms of the original transform steps $R_{max} \simeq p_{max}/N\Delta x$ where Δx (nm) is the scan step on the specimen ($= D/M$ in image steps; § 3.4) and p_{max} is an estimate of the maximum significant spatial frequency (in sample points) in the image transform. For an image scanned in a raster of 100×100 with $\Delta x = 0.5$ nm, $R_{max} = 0.5$ nm^{-1} corresponds to $p_{max} = 25$ transform steps. However, in most applications R_{max} should be significantly smaller than 0.5 nm^{-1}; a power spectrum is then first produced using the full transform ($p_{max} \simeq N/2$) before a decision is made on the choice of R_{max} (or its equivalent in transform steps). The transform $F(R, \Phi)$ is symmetrised as above, and the coefficients $G_n(R)$ are calculated using the inverse relationship of Eq. (4.6):

$$G_n(R) = \frac{1}{2\pi} \int_0^{2\pi} F(R, \Phi) \exp\left[-in(\Phi + \pi/2)\right] d\Phi \qquad (4.9)$$

The rotational power spectrum P_n can then be calculated from:

$$P_n = \varepsilon_n \int_0^{R_{max}} |G_n(R)|^2 \, 2\pi R \, dR \qquad (4.10)$$

Note that Eqs. (4.5) and (4.10) are equivalent because the total power in the image and in the transform must be identical (conservation of energy); this relation between the image and its transform is sometimes referred to as *Parseval's theorem*:

$$\int_0^{r_0} |g_n(r)|^2 \, 2\pi r \, dr = \int_0^{R_{max}} |G_n(R)|^2 \, 2\pi R \, dR \qquad (4.11)$$

for transform pairs $g_n(r)$ and $G_n(R)$.

From the power spectrum the maximum value of n for which P_n is significantly larger than the noise (components for which n is not a multiple of m) can be estimated.

In Fig. 4.3a the P_n value for $n = 18$ is significantly larger than the value of the adjacent noise components $n = 17$ and $n = 19$, whereas in Fig. 4.3b P_n for $n = 3$ is not significantly above the noise level for $n = 2$ and $n = 4$. The rotational power spectrum gives a quantitative measure of the degree of preservation of the rotational symmetry of the particle: the maximum value for $n \, (= $ a multiple of m) determines the maximum significant order of Bessel function that contributes to the $G_n(R)$ in Eq. (4.7). The first peak in the Bessel function $J_n(X)$ occurs at approximately $X = n + 2$, so that the value of the Fourier cut-off R_{max} is determined from:

$$X = 2\pi R_{max} r_0 \simeq n + 2 \qquad (4.12)$$

For the groups of nine hexons, $n = 18$, the radius of the particle $r_0 = 18$ nm, so that from Eq. (4.12), $R_{max} \simeq 1/5.7 \, \text{nm}^{-1}$, corresponding to $p_{max} \simeq 9$ transform steps. Thus there will be little point in using transform radii R significantly greater than R_{max} for calculating the rotationally filtered image using Eqs. (4.8) and (4.2) because these components contribute mainly as noise. For display the filtered image $\rho(r, \phi)$ is reinterpolated from polar coordinates to Cartesian coordinates.

If the rotational symmetry of the particle is not evident from the image, a rotational power spectrum is produced for a series of values of m; for each value the image transform is symmetrised to maximise the m-fold components of the power spectrum. For incorrect m values, the values of P_n corresponding to m-fold symmetry ($n = m, 2m, 3m, \ldots$) are not significantly larger than adjacent values of P_n; e.g. for a protein disc from TMV, P_{17} is significantly greater than either P_{16} or P_{18} for a 17-fold rotational centre, but neither P_{16} nor P_{18} is much larger than P_{17} when either a 16-fold or 18-fold rotational centre is chosen (Crowther and Amos 1971). It is also possible to prove whether or not a particle has any significant rotational symmetry, when visually it appears to do so. The power spectrum allows the preservation of a particle to be assessed in two ways. Firstly from the highest significant angular harmonic n it is possible to say to what resolution the rotational symmetry is preserved. Secondly the fraction of the total spectrum P that corresponds to m-fold rotational symmetry gives an overall measure of the symmetry of the particle.

4.1.3 Computer methods: applications

Rotational filtering has been applied to several biological assemblies such as TMV and T4 base plates (Crowther and Amos 1971), groups of nine hexons from adenovirus (Crowther and Franklin 1972) and alfalfa mosaic virus (Mellema and van den Berg 1974) and can be applied to the enhancement of the image of any particle which shows evidence of m-fold rotational symmetry.

Rotational filtering alone will not necessarily reveal biologically significant information. The signal-to-noise improvement is only a factor of $\sqrt{m}$ so that it is unlikely that the technique can be applied to unstained specimens and low exposure images with a poor signal-to-noise ratio. The technique is likely to be applied to images of negatively-stained specimens, which will never show very fine structural detail. However, if it can be combined with a biochemical study it may yield important information. A good example is the study of the T4 base plate (Crowther 1976; Crowther et al. 1977). This is composed of various proteins (gene products) that change their arrangement (conformation) when the phage attaches to a bacterium. Biologically it is required to determine which gene products are essential to the binding of the phage and to the penetration of the bacterial cell wall. The inactive conformation of the T4 base plate (extended tail) is hexagonal. This changes to a star conformation on binding to the bacterial cell wall (contracted tail). The locations of the various gene products in the base plate are not at all evident in the electron microscope images (Fig. 4.5). However, 6-fold rotational filtering of a base plate allows a comparison to be made between the conformations of the hexagonal and the star forms. Rotational filtering has been combined with a biochemical study with T4 mutants that omit some gene products from the base plate assembly (Crowther et al. 1977). The maximum value of n which is significant in the power spectrum of the T4 base plate is about 30, corresponding to $R_{max} \simeq 1/4.1 \text{ nm}^{-1}$ ($r_0 \simeq 21 \text{ nm}$); even with this relatively poor resolution it is possible to see the large changes in configuration that occur in the T4 base plate system, and to determine the location of the gene products in the base plates.

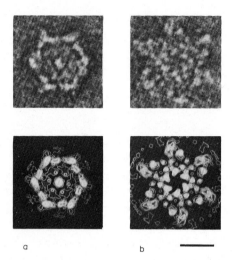

a   b

Fig. 4.5.   Rotational filtering applied to the negatively-stained base plate of T4 bacteriophage, (a) hexagonal, (b) star configuration, together with their 6-fold rotationally averaged images. Contours denote high density of 'protein'. The star configuration arises on binding the phage tail to bacterial cell walls. Even though the rotationally filtered images display the structural features much more clearly than the original micrographs, interpretation of the images is only possible by comparison with mutant structures lacking known gene products. Image bar = 20 nm. (From Crowther 1976.)

4.1.4   Photographic versus computer methods

The use of photographic superposition for rotational averaging has the advantage of simplicity, but in other respects it is inferior to numerical methods. The numerical analysis of rotationally symmetrical structures allows a quantitative assessment to be made of the preservation of the particle, and also serves to distinguish between particles with apparent and real rotational symmetry. Often a particle will look as if it has rotational symmetry but an objective examination of the power spectrum shows that there is no significant m-fold component. Photographically it is always possible to produce a rotationally filtered image but it is difficult to tell whether the final result contains any genuine information. The highest significant angular harmonic from the numerical transform shows to what resolution $(1/R_{max})$ the structure exhibits m-fold rotational symmetry; structural detail finer than $1/R_{max}$ will by definition show m-fold symmetry but it may arise only as a result of symmetrising noise components. The computer method thus allows non-structural information to be eliminated from the rotationally filtered result by setting maximum values for n and R; photographically some fine detail, which may be false, will be present in the image, even after m-fold averaging.

If a choice has to be made of several possible values for m the computer method has the advantage that using the same set of numerical data it is only necessary to produce the required power spectra for a series of m values. Photographically a rotationally filtered image for each value of m will have to be produced separately, followed by a visual inspection in order to choose m.

In the computer it is relatively easy to add together or to compare the results of rotationally filtering the images of several different particles. Information from k particles may be added together to give a further improvement in signal-to-noise of $\sqrt{k}$, provided that the relative orientations of the particles are determined from the display of the rotationally filtered images. In calculating $\rho(r, \phi)$ from Eq. (4.2) an angle ϕ_0, corresponding to the relative rotation of each particle, is added to ϕ. The density values for each particle can then be added together because the rotational centre of each image has the same origin of coordinates.

4.2 Specimens with translational symmetry

The image of a specimen with one- or two-dimensional order can be enhanced by *translational averaging*; that is, superimposing nominally identical units by translating the image by the repeat distance of the structure. Alternatively, using the image transform, the equivalent procedure is referred to as *translational filtering*, because only those parts of the transform that result from the periodic nature of the image are allowed to contribute to the reconstructed image. In contrast to rotational filtering, translational filtering can be implemented optically as well as numerically; in the former a mask is made with holes corresponding to the diffraction maxima in the optical transform and with this mask aligned on the optical bench, the filtered image is produced by means of an imaging lens (Horne and Markham 1972). In the computer equivalent, all parts of the numerical transform not consistent with the periodic structure are set to zero and an inverse Fourier transform produces a filtered image. If neither optical nor computing facilities are available, a translationally averaged image can be produced photographically using an enlarger.

4.2.1   Photographic superposition

Each unit in the periodic structure can be considered to comprise the motif (signal, $s(r)$) with noise $n'(r)$ superimposed. The noise, n', may arise from several sources including non-uniformity of staining, radiation damage, and the carbon support film. The image intensity, $j(r)$, in the unit can then be written as:

$$j(r) = s(r) + n'(r) \qquad (4.13)$$

By adding k units the signal $s(r)$ is increased by a factor k, whereas the noise $n'(r)$, which is non-periodic, will increase only by $\sqrt{k}$ (assuming n' is random, and not structure correlated). Thus by superimposing k units the signal-to-noise ratio, S/N, can be increased by $\sqrt{k}$.

Markham averaging (Horne and Markham 1972) is achieved by photographically superimposing nominally identical motifs. The best images are selected from their optical diffraction patterns, using only images with no contrast transfer reversals or transfer gaps up to the highest resolution diffraction spots observed. Photographic superposition of one-dimensional structures is relatively easy, but two-dimensional photographic averaging has to be made in two steps, corresponding to two sets of one-dimensional superpositions along the principal axes of the lattice.

The image is projected at a magnification of 10–$20 \times$ using a photographic enlarger so that the periodicity is about 5–10 mm, and the distance d corresponding to p repeat units along one axis of the periodic structure is measured; the repeat distance of the structure d/p is then calculated. The superpositions are achieved by translating a sheet of photographic paper or film (for two-dimensional averaging) by d/p and making k exposures each for an exposure time of $1/k$th of the normal exposure time for making a single print. In order to achieve accurate translational alignment the table from a standard Joyce–Loebl microdensitometer can be used. The traverse of the table is aligned as accurately as possible with the axis of the periodic structure, and then the photographic film or paper is fixed firmly to the table and covered with card. After each translation of d/p the card is removed and an exposure is made. Figure 4.6a shows the image of a negatively-stained crystal of adenovirus fibres together with the photographically averaged result (Fig. 4.6b) after 5 lateral superpositions in one direction; note that the repeat distance in the horizontal direction is twice the apparent repeat because the fibres are arranged in an alternating pattern. From the image

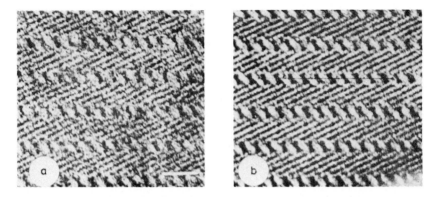

Fig. 4.6. (a) Bright-field image of a negatively-stained fibre crystal from adenovirus, (b) the result of five photographic superpositions. Image bar = 20 nm. (V.M. Mautner and N.G. Wrigley, unpublished.)

alone this actual repeat distance is not easy to see, so the optical diffraction pattern must be first used to determine the actual periodicity of the structure. If the periodicity determined from the optical diffraction pattern is a then the translational movement of the densitometer table should be Ma, where M is the *total* magnification of the projected image. Alternatively a trial and error procedure can be used where separate photographic superpositions are made for multiples of the apparent repeat distance. Only superpositions made for the correct period will have a similar appearance; thus for the example shown in Fig. 4.6 the superpositions arising from 2,4,6 etc. apparent repeats will give similar results, and will differ from the results for 1, 3 and 5 apparent repeats. The maximum number of superpositions that can be achieved photographically is in the region 5–10 because of the problem of accurately aligning the densitometer table with the axis of the periodic structure. Clearly photographic superposition is only possible with images where the repeat structure can be observed with fairly high contrast and so the technique is probably limited to stained specimens; this is not a limitation with Fourier filtering because the periodic information is concentrated at diffraction maxima.

For a two-dimensional superposition the negative from the first one-dimensional averaging is placed in the enlarger and projected at unit magnification. The repeat distance along the second axis of the structure is determined and superpositions onto film or paper are made by translating the densitometer table along this direction. If the lattice is orthogonal, this second axis is at right angles to the first axis, whereas for hexagonal lattices, the superpositions are made at 60° to the first set of superpositions. Usually distortions in the lattice limit the total number of superpositions that can be achieved on a two-dimensional lattice to about $5 \times 5 = 25$; superimposing units that are becoming increasingly out of register as a result of lattice distortions produces a poor result with little internal structure in each motif. This limitation applies equally well to Fourier filtering but in this case corrections may be applied for small lattice distortions if computer processing is used (§ 4.2.3).

4.2.2 Computer and optical methods: analysis

Images with good quality optical transforms are scanned with the direction of the scan aligned as well as possible with one principal axis of the periodic structure. Badly aligned images may produce relatively poor numerical transforms. For orthogonal lattices it is possible to produce a digitised image with sample points aligned with both axes, but with non-orthogonal lattices (e.g. hexagonal lattices) ideal sampling can only be made along one principal axis. The numerical transform may show blurred diffraction maxima as a result of imperfect sampling of the image, in addition to the blurring arising from genuine lattice distortions; only a perfect lattice with identical subunits produces a transform with sharp diffraction maxima. Distortions arise from short-range effects, such as the relative rotation or different staining pattern of nominally identical subunits, as well as long-range effects such as curvature of the lattice. Ideal sampling of the image would correspond to scanning along the principal axes of the lattice, and an image where there is an exact (integral) number of sample points in a repeat unit and an exact number of repeats in the scan; these two requirements are a result of the properties of the discrete Fourier transform. It is often impossible to distinguish between the different causes of the broadening of diffraction maxima; however, in the computer processing of images, sampling can be improved by *interpolating* the scanned image onto a new lattice where the sampling requirements of the discrete Fourier transform can be satisfied (§ 4.2.3).

Fourier filtering depends on the fact that periodic information is concentrated at particular spatial frequencies in the reciprocal lattice, whereas the noise is distributed uniformly. Thus the image transform $J(v)$ consists of two components, the signal $S(v)$ and the noise $N(v)$:

$$J(v) = S(v) + N(v) \tag{4.14}$$

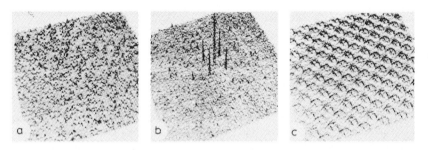

Fig. 4.7. Illustration of the averaging of a statistically noisy image: (a) image, (b) power spectrum (Fourier transform), (c) reconstructed image obtained by filtering using only periodic information. (From Kuo and Glaeser 1975.)

The principle of Fourier filtering is illustrated in Fig. 4.7 (Kuo and Glaeser 1975). Figure 4.7a shows the image of a periodic structure 'contaminated' by noise and Fig. 4.7b shows its power spectrum (the modulus squared of the Fourier transform $|J|^2$), where the periodic information is shown by the large peaks clearly separated from the noise in the spectrum. In an image reconstructed by omitting from the transform the non-periodic information, the periodic structure is clearly seen (Fig. 4.7c).

Having produced a numerical transform, the lattice should be indexed (§ 3.7) before the transform is filtered; an incorrect indexation of the lattice can lead to choosing the incorrect Fourier coefficients for the reconstruction with the possibility of producing an image with false detail (Taylor and Ranniko 1974).

The filtering operation corresponds to multiplying the image transform, Eq. (4.14), by a *mask* $M'(v)$ which is unity at or near reciprocal lattice points and zero elsewhere in the transform; $M'(v)$ can be represented as the convolution (∗) of a reciprocal lattice function $L(v) = \sum \delta(v - ha^* - kb^*)$ (Eq. (3.10); § 3.1.2) with a *window function* (the mask hole) $W(v)$; that is, the image transform is modified to (Aebi et al. 1973):

$$J(v) \times [L(v) * W(v)] \qquad (4.15)$$

Ideally $W(v)$ should correspond to one sample point in the numerical transform, because this corresponds to averaging over the whole image (see below). However, as a result of imperfect sampling and lattice distortions, the diffraction maxima will spread over several transform points. So $W(v)$ is chosen to include as many significant diffraction maxima as possible and usually $W(v)$ corresponds to 3×3 transform points, and this corresponds to averaging over a limited number of periodic units. Thus the choice of the mask size is equivalent to determining the amount of averaging in the original image. An inverse transform of Eq. (4.15) gives the reconstructed image:

$$j(r) * [l(r) \times w(r)] \tag{4.16}$$

since a multiplication in Fourier space corresponds to a convolution in real space and vice versa (Eq. (3.22); § 3.2). This is equivalent to a weighted averaging of the original image over the lattice $l(r)$. The weighting function $w(r)$, which is the Fourier transform of $W(v)$, determines the number of units over which the averaging occurs. The weighting function in a Markham superposition is unity over the number of unit cells superimposed, but for Fourier filtering, only the unit cell at the centre of a particular array in the lattice has unit weighting and units away from this centre are added with a reduced weighting $w(r)$. Because of the inverse relation between real and Fourier space, a small mask hole $W(v)$ corresponds to a broad weighting function $w(r)$ and averaging over a large number of unit cells. Conversely a large mask hole corresponds to a narrow weighting function with averaging over only a few unit cells about a particular centre in the image. This applies equally well to optical filtering where the mask is a physical one and the size of the holes in the mask determines the amount of local averaging. In order to determine the amount of averaging, the weighting function $w(r)$ must be calculated from a knowledge of the mask size used. There is a choice of square or circular mask holes, the latter being the more convenient for producing optical masks. For a square hole of size d, expressed in terms of reciprocal lattice constant units, the weighting function at a distance (a_x, a_y) from a 'reference' unit cell, is:

$$w(a_x, a_y) = \frac{\sin(\pi d a_x) \sin(\pi d a_y)}{\pi^2 d^2 a_x a_y} \tag{4.17}$$

where the distances a_x and a_y from the reference unit cell to its neighbours are also expressed in lattice constant units; that is, in terms of the number of unit cells from the reference. Thus for $a_x = 0$ and $a_y = 0$, $w = 1$, corresponding to unit weighting of the reference cell, and as a_x and a_y increase, w gradually decreases. If d is chosen to be 1/10th of the reciprocal lattice distance and $a_x = 5$ ($a_y = 0$), the weighting function $w = \sin(\pi/2)/(\pi/2) = 0.67$. Thus the unit cell 5 repeats away from the reference unit is added with a weighting of 0.67; further out from the reference unit, the contribution from unit cells to the averaged result is further reduced, so that for $a_x = 10$ unit cells the contribution is negligible (sin $(\pi d a_x) = 0$). Using a square mask hole with a size $d = 1/10$th of the reciprocal lattice distance, averaging occurs over a maximum of 10×10 repeat units, but most of these units are added (superimposed) with a weighting of less than 0.5. Thus Fourier filtering is equivalent to a Markham averaging with a reduced contribution from units that are some distance from the reference unit cell. For a circular mask hole the weighting function w is rotationally symmetric:

$$w(a) = \frac{2}{\pi a d} J_1 (\pi a d) \qquad (4.18)$$

where a is the distance from the reference unit cell in terms of lattice constant units and J_1 is the Bessel function of order one; the first zero of J_1 corresponds to $\pi a d = 3.83$, so that for values of $\pi a d$ larger than 3.83, unit cells do not contribute significantly to the averaged result. The reconstructed image of the neuraminidase crystal shown in Fig. 3.30b (§ 3.7.1) was produced from a numerical transform using a circular mask hole of 'diameter' 5 sample points, and the reciprocal lattice distance corresponded to 20 sample points; thus $d = 5/20$. Then $w(a)$ becomes zero when $\pi a \times 5/20 = 3.83$ or $a \simeq 5$, and only unit cells 4 repeat distances away from the reference cell contribute to the averaged result, and the averaging in two-dimensions is over approximately 4×4 unit cells, with a consequent increase in the signal-to-noise ratio of approximately $\sqrt{16} = 4$.

Although a larger signal-to-noise enhancement can be achieved by using small mask holes, such holes are often undesirable for distorted lattices because contributions from unit cells increasingly out of register will be added to the reference unit cell. Averaging should only be produced over an area of image for which the units are aligned on the perfect lattice to a resolution significantly smaller than the size of the unit structure; superimposing units significantly out of register only produces blurring. The value for d usually chosen is between 1/5th and 1/10th of the reciprocal lattice distance.

An approximate calculation of the number of unit cells over which averaging occurs can be made from the number of unit cells in the original image scan and the size of the mask hole in transform sample points. A scan corresponding to $r \times r$ unit cells with a square mask hole size $m \times m$ sample points corresponds approximately to averaging over $(r/m) \times (r/m)$ unit cells. Thus with $m = 1$ averaging occurs over the whole area of the scanned image.

For the neuraminidase image shown in Fig. 3.30 the image area corresponded to 195 nm × 195 nm (256 × 256 scan of an image of magnification 52,000 with a scan spot size 40 μm) or 20 × 20 unit cells. Thus using a mask hole size corresponding to 5 × 5 sample points, averaging occurs over approximately 20/5 × 20/5 $\simeq$ 16 unit cells.

The reconstructed image can be displayed on a television system, or as overprinted characters on a line-printer, or as a contour map on a graph plotting device.

Optical reconstructions can be made using copper masks produced by photo-etching. Usually the optical diffraction pattern is photographically enlarged onto film to a convenient size for punching mask holes of the correct size. If holes are punched with a cork-borer of 2 mm diameter, the optical diffraction pattern should be enlarged so that the reciprocal lattice constant corresponds to the chosen value of d. For example, if $d = 1/5$, then the diffraction pattern should be enlarged so that the reciprocal lattice distance is 10 mm. The enlarged diffraction pattern is then indexed and the reciprocal lattice scratched onto the emulsion side of the enlargement. The film is than placed on black paper and holes punched through the film onto the black paper to produce a mask *template*. The template is then photographed onto lithographic film using the same demagnification as the original enlargement of the optical diffraction pattern; this is best achieved by placing the original optical diffraction pattern in the enlarger and adjusting the enlarger so that the holes in the template coincide with the projected image of the diffraction pattern. Unexposed film then replaces the diffraction pattern and the film is exposed to the template on a light box to produce a high contrast copy of the reciprocal lattice. This demagnified mask template is now copied onto a copper foil (thickness about 50 μm) coated with a negative photoresist by exposure to ultraviolet light. The copper foil is etched using a solution of $FeCl_3/HCl$ to produce a copper mask with holes which should match exactly the diffraction maxima in the optical transform. The mask is aligned in the diffraction plane of an optical bench suitable for optical reconstruction (Horne and Markham 1972) so that with the image in place, the optical transform and mask holes are correctly aligned. Usually the reconstructed image produced in such a way has a low contrast because of the large uniform background due to the zero-order laser beam. Frequently the zero-order beam is attenuated by placing electron microscope grids over the central hole of the mask; the attenuation is often as high as 80%. However, care must be taken in attenuating the zero-order beam because it also acts as a reference wave for interference with the diffracted waves which produce the reconstructed image. If the zero-order beam is attenuated too strongly, frequency doubling can occur and the reconstructed image will show half-spacings, as can occur similarly in dark-field microscopy (§ 5.4). A series of reconstructed images should be taken using a range of zero-order attenuations, and the highest contrast reconstruction that shows no evidence of half-spacings chosen. This problem does not occur in computer reconstruction because the amplitudes and phases of the diffraction maxima are determined numerically and no reference wave is required to 'preserve' this phase information. This brief description of optical filtering is not intended to be comprehensive but to complement the information given in Horne and Markham (1972) and to provide a comparison with computer techniques.

4.2.3 *Computer methods: refinements and results*

The main refinement of computer methods of translational filtering is the production of an ideally sampled image by *interpolation*; the first step is to produce a *coarse* filtered result using relatively large 'mask' holes, so that the averaging is localised over small areas of the image. This is sufficient to produce an image on the line-printer where the main features of the lattice are recognisable. The two principal axes of the lattice are then drawn on the line-printer output; these may not correspond to the actual orthogonal axes of the original scan. Thus Fig. 4.8 shows the scan axes (original axes) and the axes (new axes) that correspond to the principal axes of the lattice. The process of determining the new image density values for this new coordinate system from the original density array is called interpolation.

For most interpolation computer programs it is sufficient to define the new axes in terms of the original coordinate system and to specify the new sampling distance (dx, dy) in terms of the original image sampling $(\delta x, \delta y)$. The density values j in the new coordinate system are determined by *bi-linear interpolation* from the values of the four adjacent density values j_1, j_2, j_3, j_4 in the original image (Aebi et al. 1973). Thus if x and y represent the distance of j from j_1 in terms of *sample points*, then the interpolated value of j is:

$$j = (1 - y)\left[(1 - x)j_1 + xj_2\right] + y\left[(1 - x)j_3 + xj_4\right] \qquad (4.19)$$

Thus if the new image point happens to coincide with j_1, $x = y = 0$ and $j = j_1$, whereas if j occurs exactly at the centre of the four nearest points $(x = 0.5, y = 0.5)$ $j = 0.25(j_1 + j_2 + j_3 + j_4)$.

As an example we can consider how the image of a neuraminidase crystal

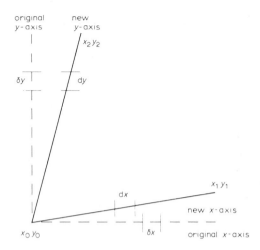

Fig. 4.8.   Interpolation of an original image onto a new (undistorted) lattice with change in sampling $(\delta x, \delta y)$ to (dx, dy) to obtain correct sampling for the discrete Fourier transform (FFT).

was interpolated onto a new coordinate system so that the axes are exactly aligned with the principal axis of the lattice, and so that there are an integral number of sample points in each repeat and an integral number of repeat units along each axis. The original origin of coordinates is defined as the sample point $(1, 1)$ on the line-printer display, whilst the original x- and y-axes are defined by sample points $(256,1)$ and $(1,256)$, respectively. The new origin is chosen to be at the centre of a unit cell at sample point $(3, 4)$, whilst the new x- and y-axes drawn on the line-printer display are slightly inclined to the original axes and are defined by sample points $(240, 6)$ and $(5, 240)$, respectively. Then the number of sample points in say 10 unit cells is determined by counting the number of sample points on the line-printer display; for the neuraminidase lattice this corresponds to 129 sample points in both directions (the lattice is square), so that the number of sample points/unit cell is 12.9; this is now chosen to be 16 sample points/unit cell so that the ratio of the new sampling to the old sampling distance, the *scaling factor*, $= 12.9/16 = 0.806$. An exact number of unit cells (16) in each direction can be obtained by choosing $NX = NY = 256$, so that the image array consists of 16×16 units compared to the original scan area of 20×20 units. This loss of about half the units in the image is unimportant because averaging is seldom made over an area larger than about 50 units. The numerical transform of this interpolated image is usually a significant improvement on the original image transform and any blurring of the diffraction maxima is now solely as a result of genuine distortions of the lattice. The filtered result is now produced from this refined transform.

If the lattice constants in the two directions are not the same, different values for the scaling factors $dx/\delta x$ and $dy/\delta y$ are used so that the sampling is perfect in both the x- and the y-directions. For non-orthogonal lattices, such as hexagonal lattices, it is, in principle, possible to interpolate the lattice from one where the axes are inclined at $60°$ to one where the axes are orthogonal. However, this is rather a large interpolation procedure using only a simple bi-linear interpolation and it may be preferable to define a new lattice that corresponds to two units of the original hexagonal lattice giving a rectangular '*super-lattice*' (Aebi et al. 1976); interpolation is then based on this *super-lattice*, to give an image with an integral number of superunit cells in each direction and with each superunit cell sampled with an even number of sample points in each direction. The image transform is filtered by setting to zero all transform coefficients not falling on the reciprocal lattice of the superlattice and setting to zero every alternate spatial frequency in both directions; the latter corresponds to the zeroing of those super-reciprocal lattice points that are not also reciprocal lattice points of the hexagonal lattice. After an inverse transform, a filtered lattice is obtained that is nearly hexagonal, differing only by a small scaling factor in the direction of the superlattice vector.

The second refinement of the computer method is to symmetrise the lattice, setting symmetry related amplitudes in the transform to the same values, usually the average value of these amplitudes. Thus for the neuraminidase (lattice plane group p4mm) all 4-fold related amplitudes are made equal (see Fig. 3.22 in § 3.4.3) and all phases set to either 0 or π (see Table 3.3 in § 3.4.3).

Interpolation can also be used to correct small distortions in the lattice. Such corrections have been made to a deformed helical filament (Fraser et al. 1976) and distorted surface layers from two species of *Clostridia* (Crowther and Sleytr 1977). Figure 4.9a shows the image of a single layer from *Clostridium thermohydrosulfuricum* together with the optical diffraction pattern from the boxed area (Fig. 4.9b). A computer reconstruction of the area in Fig. 4.9c is shown in Fig. 4.10a; no substructure is evident as a result of both short- and long-range disorder of the hexagonal subunits. In order to correct for these distortions the disordered hexagonal lattice is effectively interpolated onto a perfect hexagonal lattice. This first step is to produce a coarse filtered result (Fig. 4.9d) using only the first order diffraction maxima and large mask holes. This gives just the overlying distorted lattice part of which

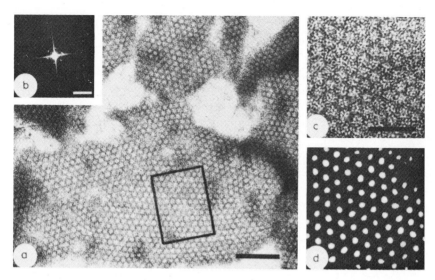

Fig. 4.9.   (a) An area of negatively-stained hexagonal S-layer from *Clostridium thermohydro-sulfuricum* (image bar = 100 nm). In this and subsequent figures the stain is dark, so that by implication protein appears light. (b) The optical transform is taken from the boxed area (diffraction bar = $0.2 \, \text{nm}^{-1}$). (c) Shows an enlarged and rotated view of the boxed area in (a) (image bar = 50 nm) and (d) is the lattice image of the area shown in (c), produced by computer filtering using only the lowest order peaks in the transform and large apertures in the filter mask. Notice that the place where the lattice image becomes very weak in the top right-hand corner corresponds to an area of S-layer in which the pattern of morphological units disappears. (From Crowther and Sleytr 1977.)

is shown in Fig. 4.11b. An undistorted average unit cell (Fig. 4.11a) can be produced from the distorted lattice (Fig. 4.11b) by the addition of corresponding density values in each distorted unit cell. Alternatively the whole image array can be interpolated to remove lattice distortions (based on the coarse filtered result) and then a Fourier transform calculated. By comparing the amplitudes of the Fourier transforms of the original and interpolated images, it is possible to assess the effects of interpolation, since the peaks in the transform corresponding to the periodic part should now contain more power after interpolation, and may be detectable above the background noise to a higher resolution. In the example shown in Fig. 4.10 interpolation (Fig. 4.10b) leads to a 40 % greater power recovery compared with simple averaging (Fig. 4.10a) without regard to the lattice distortions; it is evident that Fig. 4.10b shows sharper (higher resolution) features and better 6-fold symmetry than does Fig. 4.10a. A further improvement in the signal-to-noise ratio can be achieved by performing a 6-fold rotational averaging of the unit cell.

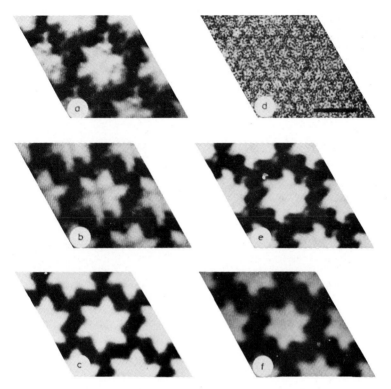

Fig. 4.10.   (a), (b), (c) show averaged images formed from part of the area shown in Fig. 4.9(c). In each case an area of 2 × 2 unit cells is shown, so that the edge of each figure corresponds to a length of 29.2 nm. In (a) the averaging is done without correcting the distortions, while in (b) the distortions were corrected. (c) is the 6-fold rotational average of the image in (b). (d) shows another area of negatively-stained hexagonal S-layer (image bar = 50 nm) with the corresponding (e) translationally and (f) rotationally averaged images. (From Crowther and Sleytr 1977.)

Figures 4.10 (d–f) show another example of the translationally (e) and rotationally (f) averaged images for a different specimen area. The two sets of processed results in Fig. 4.10 show similar features verifying that the interpolation procedure has produced consistent, and therefore genuine, structural features; there is, for example, evidence that each point of the hexagonal 'rosette' is connected to neighbouring rosettes by a network of fine bridges. It is always important to produce reconstructions from several images so that structural consistency can be checked from one reconstructed image to the next; a reconstruction from a single image cannot necessarily be believed – it may be atypical.

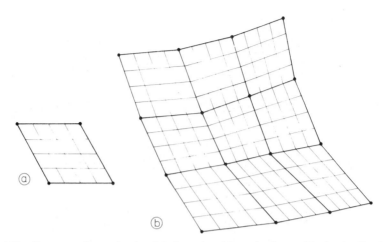

Fig. 4.11.   How an undistorted unit cell (a) is produced from the distorted lattice (b). The dots represent the centre of the repeating motif, as determined from the lattice image, and the heavy lines represent the unit cell edges. Each cell is divided into a number of equal parts corresponding to the sampling chosen for the undistorted cell. By joining corresponding pairs of points on opposite cell edges a distorted sampling grid (light lines) is generated in each distorted unit cell. The density values at these grid points can now be interpolated from the underlying array of densitometer values and added into the average cell. (From Crowther and Sleytr 1977.)

Figure 4.12a shows the image of a single protein layer from the envelope of *Escherichia coli*, together with its optical transform (Fig. 4.12b) (Steven et al. 1977). The computer filtered result (Fig. 4.12c) is obtained after scaling (interpolating) the original image so as to maximise the power in the Fourier coefficients of the indexed reciprocal lattice out to spatial frequencies with visible diffraction maxima on the corresponding optical transform. The symmetry of the lattice is P3 with a lattice constant of 7.7 nm; the triplet indentations (shown as black in Fig. 4.12c) penetrable by negative stain have a centre-to-centre spacing of about 3.0 nm; light areas of the image cor-respond to stain-excluding regions of matrix protein.

If the procedure given above for producing an ideal image transform by interpolation seems to be too complex, there is an alternative, simpler procedure which partially compensates for non-ideal sampling and small lattice distortions. The amplitudes around each diffraction maximum are added together (integrated). An average phase is calculated; provided that the phase origin has been determined, the phase variations near the centre of a diffraction maximum should be small enough for this procedure to be valid. These modified amplitude and phase values are calculated for each diffraction maximum and each is placed at a single reciprocal lattice point;

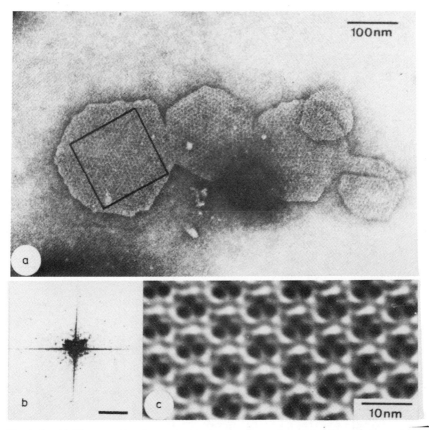

Fig. 4.12. (a) Electron micrograph of periodic arrays of negatively-stained 'matrix' protein derived from spheroplasted *E. coli* by differential treatment in sodium dodecylsulphate (SDS). (b) The optical diffraction pattern of the windowed area is indexed on a simple hexagonal lattice (only a single layer) with diffraction maxima extending to $1/2.2$ nm^{-1} (diffraction bar = 0.2 nm^{-1}). (c) Digitally filtered result in which the 'protein' is white. (From Steven et al. 1977.)

the remainder of the transform is set to zero. This procedure does produce a sharper Fourier spectrum than that originally obtained from the image, but it is difficult to say how much compensation there is for imperfect sampling and lattice distortions. Figure 4.13c is a reconstructed image obtained using this procedure for an image (Fig. 4.13a) of a negatively-stained sheet of microtubules (Erickson 1974).

Translational filtering can also be applied to images of metal-shadowed specimens, although interpretation of such images should be made with care since the appearance of the image depends on the geometry chosen for the

metal shadowing. Metal shadowing is most frequently used to examine the surface topography of biological specimens after freeze-fracturing or freeze-etching; such images have the advantage that only one surface is imaged so it is possible, for example, to observe a single layer of a membrane, whereas the conventional negatively-stained specimen may comprise a double layer (Wurtz et al. 1976; Kistler et al. 1977). For thick or bulk specimens, replication followed by metal shadowing is often the only method of obtaining high resolution topographical information. The image transform of a metal-shadowed specimen is determined by the normal procedure. It will be evident that the image transform does not show the symmetry displayed by the corresponding negatively-stained specimen. The symmetry depends on the angle of shadowing and so no attempt should be made to alter such an image transform. Depending on the direction of metal shadowing, some transform spots will be missing, because the image will show little or no topographical contrast in a direction perpendicular to the direction of the incident metal shadowing; such an image must be interpreted with care. Fig. 4.14 shows the images of metal-shadowed (a) outside and (b) inside surfaces of a T-layer cylinder from *Bacillus brevis* together with (c) the image of a negatively-stained double layer; their optical diffraction patterns are shown respectively in Figs. 4.14 (d–f) (Kistler et al. 1977). The reconstructions of the shadowed specimens are shown in Fig. 4.14g and h, together with the one-sided filtered result of the negatively-stained T-layer (Fig. 4.14i). The appearance of the reconstructed images of the shadowed specimens depends on the direction of the shadowing chosen.

With these reservations about interpreting images of metal-shadowed specimens, the method given by Smith and Kistler (1977) for correcting such images for the effect of the direction of shadow represents an important advance in image processing. Such a correction enables an image of the surface topography to be produced that is virtually independent of the angle chosen for metal shadowing. The model chosen by Smith and Kistler (1977) is illustrated in Fig. 4.15. The contrast in a shadowed specimen arises from the variation in the thickness $t(r)$ of the metal deposit on the specimen

Fig. 4.13. (a) Image of a negatively-stained flat opened sheet of a microtubule and (b) its optical transform. A thin line of stain may be seen running down the middle of each filament. The diffraction pattern shows that the periodic structure is well preserved (sharp diffraction maxima). There are two orders of the $1/5.1 \text{ nm}^{-1}$ equatorial spot and two spots on the $1/4.0$ nm^{-1} layer line. (c) Computer reconstruction based on the transform. Scales (a) 20 nm, (b) 0.2 nm^{-1}, (c) 10 nm. (From Erickson 1974.)

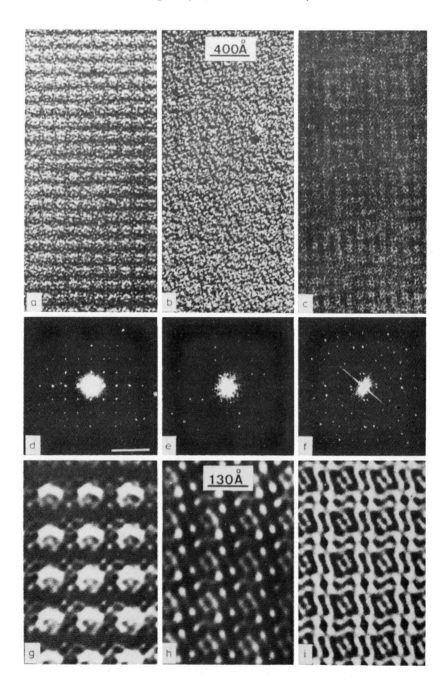

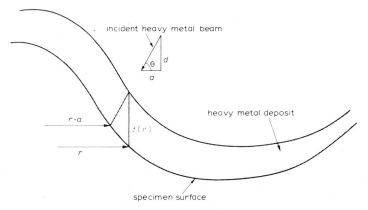

Fig. 4.15. A diagrammatic representation of the model for the deposition of heavy metal onto a specimen surface. From the direction of the incident metal beam, the metal appears to have the same depth over the whole specimen. In the vertical projection, the metal thickness $t(r)$ at a point is determined by the difference between the heights of the specimen surface and the upper surface of the metal. (From Smith and Kistler 1977.)

surface. If d represents the thickness of metal deposited on a horizontal flat surface, the thickness profile $t(r)$ is given by:

$$t(r) = d + z(r - a) - z(r) \qquad (4.20)$$

where $z(r)$ is the height of the specimen surface at r and $z(r - a)$ is the height of the surface at $(r - a)$; a is the projection of the line lying in the direction of the metal beam whose length is $d/\tan \theta$ for an angle of shadow θ. Thus the image intensity, assumed proportional to $t(r)$, consists essentially of the difference between the surface relief $z(r)$ of the specimen and a similar but displaced relief; if $\theta = 90°$, $a = 0$ and there is no contrast in the image. The highest contrast occurs for low angles of the metal shadow but such images are correspondingly further from the true surface structure. The image transform $J(v)$ is then given by the Fourier transform of Eq. (4.20):

$$J(v) = d\delta(v) + Z(v) \exp(-2\pi i v \cdot a) - Z(v) \qquad (4.21)$$

Fig. 4.14. (a) Image of freeze dried-shadowed T-layer cylinder (outside), (b) image of freeze dried-shadowed T-layer cylinder (inside), (c) negatively-stained T-layer cylinder. (d), (e), (f) are the corresponding optical transforms (diffraction bar = 0.2 nm^{-1}) and (g), (h), (i) the corresponding optically-filtered images. In the images the metal shadow is white, whilst in the negatively-stained specimen the 'protein' appears white. (From Kistler et al. 1977).

since the Fourier transform of $z(r - a)$ (the origin of z shifted by a) is the Fourier transform of $z(r)$, $Z(v)$, multiplied by the phase factor $\exp(-2\pi i v \cdot a)$.

Thus if the zero order $d\delta(v)$ is excluded from the numerical transform, the Fourier transform $Z(v)$ of the surface profile $z(r)$ can be calculated by dividing the image transform $J(v)$ by the factor $-[1 - \exp(-2\pi i v \cdot a)]$ at all spatial frequencies v for which this factor is non-zero. The image transform should be zero along the line $(v \cdot a) = 0$, since the metal shadow will not show surface variations in a direction perpendicular to the incident metal beam.

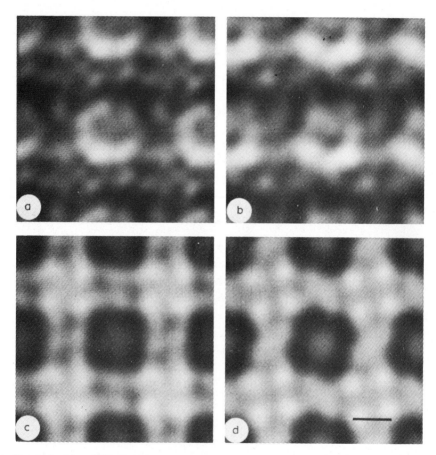

Fig. 4.16.   (a) and (b) show 2×2 computer averaged unit cells obtained from shadowed specimens of T-layer. The contrast is chosen so that the dark areas of the image indicate the absence of heavy metal deposit. (c) and (d) show 2×2 unit cells of the surface reconstructions based on the images shown in (a) and (b), respectively. Dark regions indicate high elevation. Image bar = 5 nm. (From Smith and Kistler 1977.)

This filter function has two effects: firstly it alters the phases of Fourier coefficients $J(v)$ so that they refer to the phase origin of the surface relief $z(r)$ and secondly it alters the amplitudes of $J(v)$. The restoration of the relative positions of the structural features is the most important effect of the reconstruction procedure. Figure 4.16a and b are micrographs of T-layers shadowed at angles of 60° and 45°, respectively. Their optical diffraction patterns were checked to ensure that diffraction spots perpendicular to the shadowing direction were either absent or not significantly above the background. The Fourier coefficients of the numerical transform were corrected using only the phase part of the filter function, because it was found that amplitude corrections were sensitive to imperfections in the shadowing model used (Smith and Kistler 1977). An inverse transform of the corrected Fourier

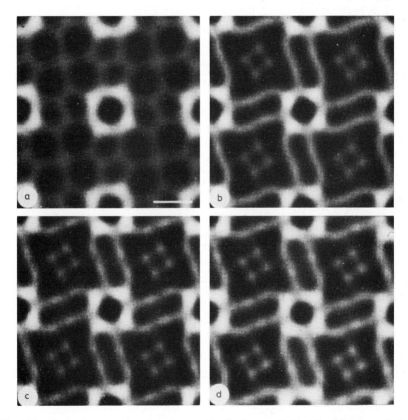

Fig. 4.17. Computer reconstructions (symmetrised) of a negatively-stained T-layer cylinder (using area A in Fig. 3.24) using (a) 3, (b) 4, (c) 5 and (d) 6 diffraction orders. Image bar = 5 nm. (P.R. Smith, unpublished.)

coefficients was then produced. The symmetry centre of the unit cell could now be determined and a 4-fold averaging performed on the corrected image. The final reconstructed images are shown in Fig. 4.16c and d; the fact that the two surface reconstructions are similar demonstrates the success of this model for metal shadowing.

A final example demonstrates the effect of increasing resolution in the image transform on the reconstructed image and emphasizes the importance of choosing only the best images for reconstruction. Figure 4.17 shows the reconstructions of images of a T-layer using an increasing number of diffraction orders (Aebi et al. 1973); the difference between reconstructions using diffraction orders up to the 3rd and 4th orders (Fig. 4.17a and b) is the most marked; the contribution from the higher diffraction orders does not seem to achieve more than a sharpening of the tetrameric structure at the centre of the unit cell. It is clear that for negatively-stained T-layers, images must display at least 4 diffraction orders (resolution = $13.1/4$ nm = 3.3 nm) before internal structure can be seen; the addition of higher resolution information does not significantly alter the interpretation of the image. It is usually informative to make several reconstructions of an image using progressively higher resolution information to seé how the appearance of the structure is affected by this additional information.

4.2.4   A comparison of computer and optical filtering

In most applications the quality of optical and computer filtered images is the same, provided some care has been taken to achieve the correct sampling (scaling) of the image in the numerical filtering procedure. Figure 4.18 shows a comparison of (a) an optically filtered and (b) a computer filtered image of T-layer (Aebi et al. 1973).

In a comparison of the two techniques the following advantages and disadvantages should be considered:

Optical filtering

1. Filtering is performed on the original photographic image

2. High-quality optics are required for the optical bench; unlike an

Computer filtering

1. A computer controlled densitometer is required to produce a digitised image

2. Fairly large computing facilities are required to handle data arrays

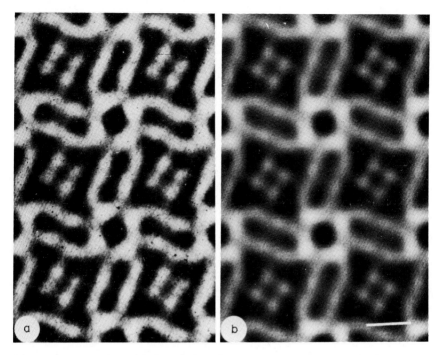

Fig. 4.18. A comparison of the (a) optical and (b) computer image reconstructions from images of a negatively-stained T-layer cylinder (unsymmetrised). Image bar = 5 nm. (From Aebi et al. 1973.)

optical diffractometer which is relatively insensitive to external vibration, the optical bench must be carefully mounted to reduce vibrations* of 256 × 256 or 512 × 512 image points, together with the associated computer programs

3. An optical transform is produced easily

3. The best numerical transforms are only produced after suitable scaling of the image

4. The masks are difficult to make and a new mask should be made for each image

4. Masking is effected numerically, and is therefore easy

* See Horne and Markham (1972) for the design and construction of optical benches

5. Image reconstruction is done photographically

5. Good display facilities are required to produce a final reconstructed image that is of comparable quality to a photograph

6. The zero-order beam has to be attenuated to produce a satisfactory reconstruction. This can lead to frequency doubling in the reconstruction

6. Because the amplitudes and phases are stored as numbers, the zero-order (background) can be reduced effectively to zero

7. The lattice cannot easily be symmetrised, except by making a subsequent Markham superposition of the reconstructed image

7. Symmetrisation can easily be achieved by numerically setting symmetry-related Fourier coefficients to be equal. Also an estimate of the errors involved in the symmetrisation can be obtained

8. Lattice distortions cannot easily be corrected

8. Lattice distortions can be corrected by interpolation of the original image onto a perfect lattice

4.3   Separating superimposed lattices

Tubular biological specimens often dry down onto the electron microscope grid to form a well preserved structure, with the two superimposed lattices rotated with respect to each other. Evidence of this type of structure shows clearly in the optical diffraction pattern as pairs of closely spaced diffraction spots. The assumption is made that the two superimposed lattices are identical, so that one set of diffraction spots arises from the top layer, whilst the second set of diffraction spots (rotated by an angle α with respect to the first set) arises from diffraction by the bottom layer (Klug and DeRosier 1966). Consequently if the image transform is masked so that only one set of diffraction maxima is allowed to contribute to the reconstructed image, a single layer can be seen. This separation of two superimposed layers cannot be achieved photographically.

The most important step in separating superimposed layers is a correct indexation of the image transform, so that the diffraction maxima are correctly assigned to one or other of the two layers (see Figs. 3.39 and 3.40

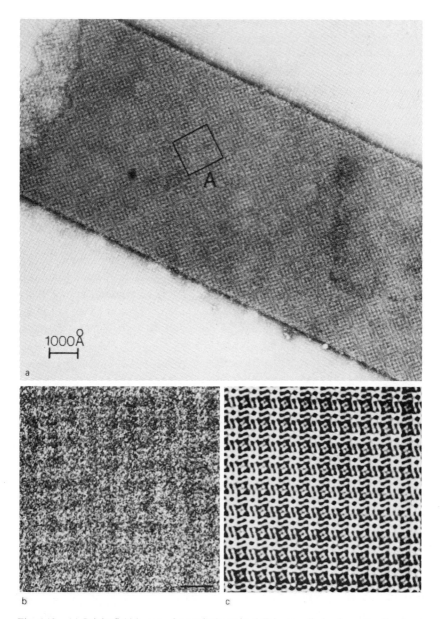

Fig. 4.19. (a) Bright-field image of negatively-stained T-layer cylinder from *Bacillus brevis*, (b) enlargement of area A (10 × 10 unit cells), (c) optical reconstruction of upper layer of area A. Image bar = 20 nm in (b) and (c). (From Aebi et al. 1973.)

in § 3.7.2). It is possible to assign a diffraction maximum to the wrong layer and to produce an image which comprises components from both layers. Figure 4.19 illustrates the application of one-sided filtering to the cylindrical form of a T-layer from *Bacillus brevis* (Aebi et al. 1973). The enlargement of the image area A (Fig. 4.19b) shows no internal structure – it is difficult to see a lattice at all, but the reconstruction of a single layer (Fig. 4.19c) clearly shows the internal arrangement of the 'protein' subunits (shown as white) in the T-layer.

The application of image processing techniques to the study of the morphogenesis of T-even bacteriophage heads is an impressive example of combining biochemistry and electron microscopy. The major capsid protein in T2 and T4 phage heads is P23*; T4 differs from T2 in having two dispensable proteins: small outer capsid protein (soc) and highly antigenic outer capsid protein (hoc) (Ishii and Yanagida 1975). The purpose of electron microscopy combined with image processing is to determine the structure and arrangement of the proteins on the surface of the phage head and to note structural alterations as the proteins hoc and soc are added to T2 to produce T4 phage heads. In order to produce large enough lattices to apply image processing giant mutants of T2 and T4 can be produced. Two such mutants are shown in Fig. 4.20a and b together with their normal forms. Figure 4.20 (c–h) show one-sided filtrations for various stages of morphogenesis, isolated by biochemical methods (Aebi et al. 1976, 1977a). The sequence (c–f) in Fig. 4.20 represents the change of a giant T2 phage head into a T4 phage head by the addition of hoc and soc proteins, changing the arrangement of the proteins (the stain exclusion pattern – shown as white). The removal of soc and hoc proteins from T4 produces a structure indistinguishable from a T2 phage head.

Figure 4.21 illustrates the next step in the study of T-even bacteriophage heads, an attempt to determine the location of the antigenic sites of the major protein P23* (Aebi et al. 1977b). Specific antibodies were raised against T2 and T4 phage heads and these antibodies were treated to produce monovalent Fab fragments, which have a molecular weight of about 50,000, and overall dimensions of $8 \times 5 \times 4$ nm. These Fab fragments have been used to stoichiometrically label the hexagonal capsid structure of T2 and T4 phage heads (Fig. 4.21a and b). By optically filtering images of areas of flattened tubular head, marked changes in the stain-exclusion pattern were observed (compare (c) and (d), and (e) and (f)). By comparing the unlabelled ((c) and (e)) with the labelled ((d) and (f)) stain-exclusion patterns of a particular periodic structure, it is possible to decide to which protein regions

Fig. 4.20. (a) and (b) show giant T2L and T4D bacteriophages, together with their wild-types. The specimens were negatively-stained with 1% uranyl acetate at pH 4.5 (c)–(h) show optical filtrations of areas containing about 150 unit cells of the near-hexagonal surface lattice of the tubular parts of various samples of giant phage. (c) Giant T2L phage with a simple 6-type capsomere morphology. Each stain-excluding region is associated with one molecule of the major capsid constituent protein P23* (mol. wt = 47,500). (d) Giant T2L complemented *in vitro* with purified hoc protein (mol. wt = 40,000) giving a (6 + 1)-type capsomere morphology with a molar ratio P23*: hoc of 6:1. The central stain-excluding pattern of each capsomere is associated with one molecule of hoc and 6 surrounding molecules of P23*. (e) Giant T2L complemented *in vitro* with purified soc protein (mol. wt = 10,000) giving a(6 + 6)-type capsomere morphology with a molar ratio P23*:soc of 6:6. The stain-excluding trimers bridging adjacent capsomeres are associated with three molecules of soc. (f) Giant T2L phage which was complemented *in vitro* with purified hoc and soc giving a (6 + 6 + 1)-type capsomere morphology with molar ratios of P23*:soc:hoc of 6:6:1. This capsomere morphology is indistinguishable from that of giant T4D phage which contains both proteins hoc and soc. (g) Giant T4D phage differentially dissociated to extract about 4 soc molecules per capsomere leaving near-equatorial soc molecules in place. (h) Giant T4D phage differentially dissociated to partially extract the soc molecules and the total extraction of the hoc molecules, still leaving some of the near-equatorial soc molecules in place. Further dissociation leads to the complete extraction of soc molecules and a 6-type capsomere morphology which is indistinguishable from that of the giant T2L phage (c). These figures demonstrate how chemical changes of a protein lattice can be correlated with structural changes by combining electron microscopy with image processing. (From Aebi et al. 1976, 1977a.)

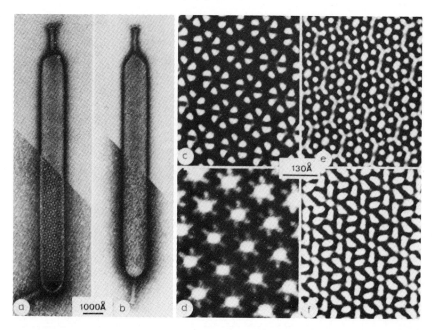

Fig. 4.21.   Electron micrographs of negatively-stained (a) giant T2L phage, (b) giant T4D phage showing the change in surface morphology by stoichiometrically labelling the major capsid protein P23* with Fab (antibody) fragments directed against a particular antigenic site of this protein: top-half *without*, bottom-half *with* antibody fragments. (c)–(f) show optical filtrations of the areas of the tubular part of the particles shown in (a) and (b), (c) T2L and (e) T4D without, (d) T2L and (f) T4D with label. The same antigenic sites seem to be located quite differently in the T2 and T4 structural states. (From Aebi et al. 1977b.)

the Fab fragments bind. In Fig. 4.21 it is evident that the same antigenic site seems to be in quite different positions in the T2 and T4 structural states.

Whilst two rotationally misaligned lattices produce two distinct sets of diffraction spots in the image transform, translationally misaligned layers give an optical diffraction pattern characteristic of a single layer (translational invariance). There may perhaps be a small rotational misalignment, giving blurred spots not necessarily clearly separated (Lake and Leonard 1974). The separation of the two lattices cannot be effected in the transform by simply dividing the amplitude by two because the phases of the transform coefficients of each layer will be different. For a lateral displacement δx between the two layers the transform coefficients from each layer will differ by a phase $\delta \omega = -2\pi \delta x \nu_x$; only for the meridional diffraction maxima ($\nu_x = 0$) when $\delta \omega = 0$ can a value for the Fourier coefficient for each layer be obtained by dividing the Fourier coefficient by 2 (assuming equal staining of both layers). Thus each transform value, amplitude F, phase Ω, will be composed of a contribution $f \exp(i\omega)$ (amplitude f, phase ω) from one layer and $f \exp[i(\omega + \delta\omega)]$ from the second layer, provided the two layers contribute equally to the image. Thus, the separation of the two layers in the image corresponds to solving in the image transform the equation:

$$f \exp(i\omega) + f \exp[i(\omega + \delta\omega)] = F \exp(i\Omega) \qquad (4.22)$$

for f and ω, knowing the transform values $F \exp(i\Omega)$ and the translational displacement of the two lattices δx. The intensity of the diffraction maximum F^2 is determined by:

$$F^2 = f^2 + f^2 + 2f^2 \cos \delta\omega \qquad (4.23)$$

Thus the intensities of the diffraction maxima in the image transform can vary from $4f^2$ for $\delta\omega = 0$ to zero for $\delta\omega = \pi$. This shows that the division of the transform amplitudes by 2 (the intensities by 4) will not produce the correct amplitudes for the transform of a single layer. From Eq. (4.23) the amplitude can be calculated using:

$$f^2 = \frac{F^2}{(2 + 2\cos \delta\omega)} \qquad (4.24)$$

It is necessary to determine the value of δx; this can only be determined if there is at least one set of diffraction maxima that is clearly separated for each layer. For a hexagonal lattice the 6-fold centre is determined for each layer using these separated diffraction orders together with a phase origin refinement procedure on the numerical transform (Lake and Leonard 1974). This procedure gives the difference δx in the phase origin between the two layers, and the corrected amplitudes f can be determined from Eq. (4.24), whilst the phases ω can be calculated using Eq. (4.22):

$$\tan \omega = \left| \frac{\sin \Omega + \sin(\Omega - \delta\omega)}{\cos \Omega + \cos(\Omega - \delta\omega)} \right| \qquad (4.25)$$

If there are no sets of non-overlapping diffraction orders it is not possible to separate the two lattices and the only course is to attempt to prepare either opened sheets of the material or tubular specimens where the two layers are rotated with respect to each other. If the two layers are not equally stained the transform amplitudes of each layer will differ, and in order to separate the two lattices, symmetry-related diffraction maxima will have to be considered (Lake and Leonard 1974).

The image of a helical structure can nominally be considered as a two-sided image and the reciprocal lattice can be separated into contributions from the near and far sides of the particle (see § 3.6.4 and Fig. 3.29). Figure 4.22 shows the image of a negatively-stained bacteriophage tail (a) together

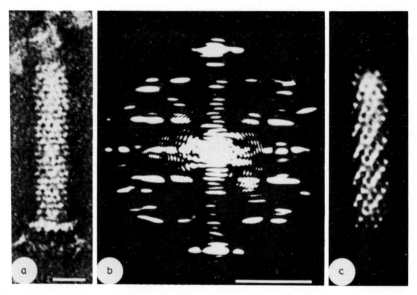

Fig. 4.22.   (a) Electron micrograph of a negatively-stained tail of T4 bacteriophage, with (b) its optical diffraction pattern. The strong meridional peak on the seventh layer line arises from the spacing of 3.8 nm between annuli. The layer lines are approximately equally spaced at $1/(7 \times 3.8)$ nm^{-1}. (c) is the optically filtered image of the tail admitting only the diffracted waves corresponding to the far side of the particle. The dominant features are two distinct sets of oblique striations, which correspond to distinct sets of helical lines running along two cylindrical surfaces of different diameter and thus correspond to features at different depths in the particle.
Image bar = 20 nm; diffraction bar = 0.2 nm^{-1}. (From DeRosier and Klug 1968.)

with its optical transform (b) (DeRosier and Klug 1968). An approximate separation of the diffraction orders is made using the procedure given in § 3.6.4 and a mask is made which transmits only one set of diffraction maxima. The reconstruction (Fig. 4.22c) shows more clearly the helical arrangement of the subunits but does not really add to the structural information.

4.4 Correction of images for defocus and lens defects

Usually the judicial choice of defocus will avoid the problem of an image with contrast reversals due to oscillations in the phase contrast transfer function sin γ (§ 3.2.2). It is sufficient to ensure that the first zero of the transfer function occurs at a higher spatial frequency than the highest resolution information in the image transform; for an ordered specimen this means that the highest order diffraction maximum should be well inside the first zero of sin γ. For negatively-stained specimens, with a resolution of at best 2 nm, defocus of less than 500 nm will ensure that this requirement is satisfied. However, as the resolution of the image improves, as for example with unstained specimens observed under low radiation dose conditions, it is difficult to take images without sign reversals and consequent transfer gaps in the image transform. In these circumstances the image should be corrected for the effects of the transfer function; this procedure is referred to as *deconvolution*. Deconvolution uses the image transform, which for a thin specimen is Eq. (3.24):

$$J(v) = \delta(v) - 2A(v) \sin \left[\gamma(v)\right] B(v) \tag{4.26}$$

Omitting $v = 0$ and spatial frequencies greater than $|v| \geqslant v_{max}$, the specimen transform $A(v)$ can be calculated by dividing the image transform $J(v)$ by $-2 \sin \gamma(v)$:

$$A(v) = \frac{J(v)}{-2 \sin \gamma(v)} \tag{4.27}$$

There are two limitations in using Eq. (4.27):
1. When sin $\gamma = 0$, there is no information in the image transform at the corresponding spatial frequencies. Consequently the reconstruction $A(v)$ will have frequency gaps corresponding to the zeros of sin γ.
2. The image transform includes a noise component $N(v)$ that has not been affected by sin γ.

Thus whenever the value of $J(v)$ is small the division by a small value of sin γ will amplify the noise components of $J(v)$. This amplification of the noise will lead to spurious structure in the reconstructed image $\eta(r)$ obtained by an inverse transform of $A(v)$. There will then be parts of the object spectrum A which cannot be reliably corrected for the effects of the transfer function. The only way to avoid this problem is to use several images recorded at different defocus values and to use only those transform coefficients where the value for sin γ is near unity. Thus a composite transform is produced from several image transforms, using the appropriate sin γ value for each image. The parameters of sin γ are usually determined from the optical transform of the carbon support film near the specimen (§ 3.3.4). For ordered specimens for which electron diffraction patterns can be obtained, a more elegant correction can be made for the effects of the transfer function by using the electron diffraction amplitudes, which are unaffected by sin γ, and the micrograph phases (Unwin and Henderson 1975) (§ 4.5). However, for non-periodic specimens there is no alternative to using Eq. (4.27) or its optical analogue.

Optical procedures for correcting for the effect of sin γ are quite complex, because a filter must be made that not only produces the amplitude variation in sin γ, $|\sin \gamma|$, but also the sign reversals in sin γ. An optical filter in the diffraction plane of the optical bench consists of two components, an amplitude filter $1/|\sin \gamma|$ and a phase filter which produces a phase shift of 0 or π according to whether sin γ is positive or negative. The details of the manufacture and use of these filters for *holographic deconvolution* are given

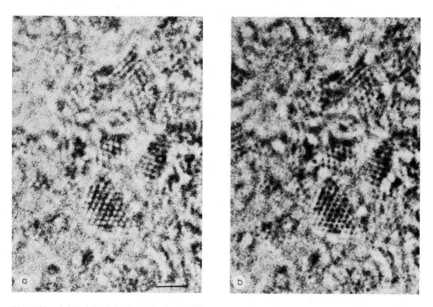

Fig. 4.23. (a) Bright-field image of a thallium crystal and the result of correction for phase contrast reversals (b). Image bar = 2 nm; $\Delta f = 120$ nm, $E_0 = 100$ keV. (From Hahn and Baumeister 1973.)

in a series of papers by Stroke et al. (Stroke 1969; Stroke and Halioua 1972; Stroke et al. 1974). More recently Burge and Scott (1975, 1976) have used *binary filters* produced by plotting on a computer microfilm facility. These filters incorporate the amplitude and phase information on a single film. Care must be taken in mounting the filters so that no spurious phase shifts are introduced into the image transform. The most important result of the correction for the effects of sin γ is the removal of the phase contrast reversals; the amplitude correction only produces alterations in the relative contributions of the Fourier coefficients to the image. Thus the simplest form of optical or computer correction corresponds to changing the phase of $J(v)$ by π each time sin γ becomes negative; numerically this corresponds to reversing the sign of the image transform in the appropriate spatial frequency ranges. For example, Fig. 4.23a shows the image of a thallium crystal recorded at a defocus of 120 nm (Hahn and Baumeister 1973). The image is then corrected optically using a $(0, \pi)$ phase filter only (Hahn 1972) to produce the reconstruction shown in Fig. 4.23b; the positions of the thallium atoms are clearly shown in the reconstruction.

4.5 The use of electron diffraction data in image processing

It is often possible to obtain high resolution diffraction data from crystalline specimens, where good quality images have been difficult or impossible to obtain, as a result of being able to record an electron diffraction pattern at a lower electron dose than that required for the corresponding image; the electrons are concentrated into visible spots in the diffraction pattern, but when spread out as an image very little can be seen. Also, as a result of the translational invariance of the electron diffraction amplitudes, specimen drift is unimportant whereas the image will be seriously affected. This is a particular problem in the examination of hydrated (Hui et al. 1974) and frozen-hydrated (Glaeser and Hobbs 1975; Chanzy et al. 1976; Taylor and Glaeser 1976) specimens where specimen stage drift limits the quality of the image. However, electron diffraction data alone cannot be used to deduce the structure of a specimen, because only the amplitudes $|A(v)|$ of the specimen transform $A(v)$ can be determined, but when combined with the information from the image transform, electron diffraction amplitudes can be used to correct for the effects of lens defects and defocus (Unwin and Henderson 1975). The image transform $J(v)$ is proportional to $A(v) \sin [\gamma(v)]$ so that:

$$J(v) \propto A_r(v) \sin [\gamma(v)] + iA_i(v) \sin [\gamma(v)] \tag{4.28}$$

where A_r and A_i are respectively the real and imaginary parts of A. The phase ω in the image transform is determined by the division of the imaginary part by the real part of J, that is:

$$\omega = \tan^{-1}(A_i \sin \gamma / A_r \sin \gamma) = \tan^{-1}(A_i / A_r) \qquad (4.29)$$

Thus the phase angle ω corresponds to the phase of $A = |A| \exp(i\omega)$, independently of $\sin \gamma$, and so values of A free from lens aberrations can be calculated by using $|A|$ from electron diffraction data and ω from a transform of the micrograph. There are two limitations: firstly at zero values of the transfer function ($\sin \gamma = 0$) there is no information in the image transform on ω and, secondly, there are ambiguities in the phase angle ω, because neither the sign of A_r nor of A_i is known. Only the sign of A_i / A_r is known, so that sign reversals of $\sin \gamma$ can give values of ω in one of two quadrants. This last problem is avoided by changing the sign of the amplitude $|A|$ whenever $\sin \gamma$ is negative. The problem of the zero values of $\sin \gamma$ is solved by the use of several micrographs at varying defocus to fill in the frequency gaps. The application of the technique to unstained purple membrane is illustrated in Figs. 4.24 to 4.26 (Unwin and Henderson 1975). The electron diffraction amplitudes determined from Fig. 4.24 are used to replace the numerical image transform amplitudes at corresponding reciprocal lattice points, whilst retaining the image transform phases. Two images are normally recorded: the first at a subminimal radiation dose to preserve the structural integrity of the specimen and the second, at normal electron dose, to determine the $\sin \gamma$ profile. Figure 4.25 shows a quadrant of the optical transform of the images taken under these conditions; Fig. 4.25a shows the diffraction maxima due to the hexagonal lattice extending to a resolution of about 0.7 nm, whilst Fig. 4.25b shows only the transform of the carbon support film. Thus in the numerical transform the sign of $|A|$ can be altered according to the sign of $\sin \gamma$. The regions of the transform where transfer gaps occur (light parts of the optical transform in Fig. 4.25b) can be combined with the transform of a second micrograph with frequency gaps in different positions. This procedure of combining electron diffraction and micrograph data cannot be achieved optically. The reconstruction resulting from combining several micrographs, together with the symmetrisation of the image transform for a P3 lattice, is shown in Fig. 4.26 in the form of a contour map (Unwin and Henderson 1975). The dense contours correspond to high concentrations of scattering material (membrane protein) whilst low density regions indicated by thinner lines are due to membrane lipid. The near

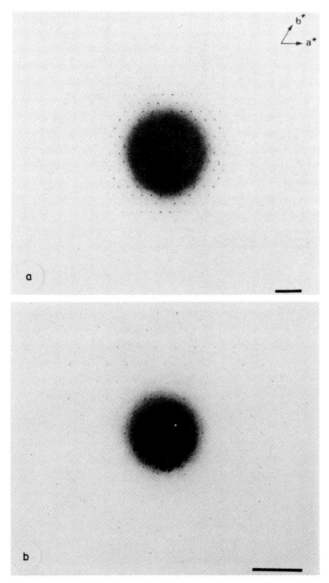

Fig. 4.24. Electron diffraction from unstained purple membrane: (a) high-angle, (b) low-angle patterns. Diffraction bar = $0.5\,\text{nm}^{-1}$. (From Unwin and Henderson 1975.)

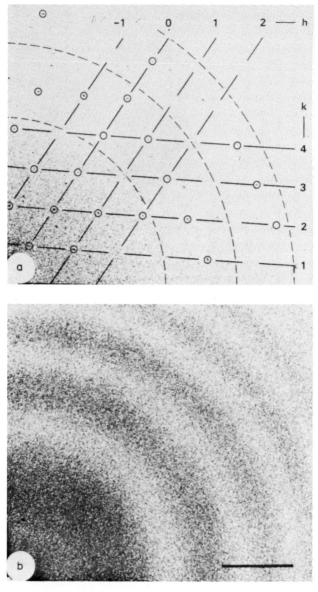

Fig. 4.25.   Optical transform quadrants of unstained purple membrane micrographs taken in bright-field microscopy: (a) low dose image, (b) high dose image. Diffraction bar $= 0.3\,\mathrm{nm}^{-1}$. (From Unwin and Henderson 1975.)

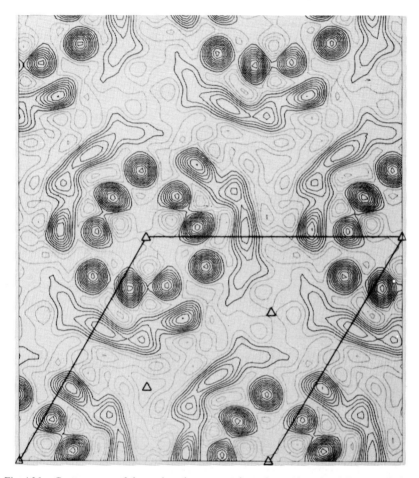

Fig. 4.26. Contour map of the projected structure of purple membrane at 0.7 nm resolution. Positive contours are shown by thicker lines; positive peaks are due to high concentrations of scattering material (protein). Low density regions indicated by thinner lines are due to lipid and glucose. Unit cell dimensions are 6.2 nm × 6.2 nm. (From Unwin and Henderson 1975.)

circular structures correspond to α-helices viewed in vertical projection, whereas the elongated structures correspond to α-helices which are inclined to the vertical. The interpretation of the projection shown in Fig. 4.26 was aided by X-ray diffraction results from oriented specimens (Blaurock 1975; Henderson 1975).

4.6   Computer processing of images of thick specimens

Specimens thicker than the limits specified for the weak phase approximation (20 nm for unstained specimens, 5 nm for stained specimens, see § 3.2.1) may not satisfy Eq. (3.23) and the image transform may have significant contributions from terms in $\cos \gamma$. In practice it is difficult to specify a maximum value for the specimen thickness for which the weak phase approximation can be used, because its validity will depend on the magnitude of the potential variations $V(r, z)$, in the specimen. The basic test of the validity of the weak phase approximation is to plot values of the image transform $J(v)$ for a particular spatial frequency v against values of $\sin \gamma(v)$ from a focus series. This test demonstrated, for example, that unstained ribosome crystals of thickness about 60 nm satisfy the approximation (§ 3.4.3; Fig. 3.21; P.N.T. Unwin, personal communication), whereas there is experimental evidence that this simple approximation breaks down for negatively-stained catalase crystals of thickness 20 nm (Erickson and Klug 1971). An extension of the weak phase approximation, allowing for relatively large phase shifts $\eta(r)$ and also taking account of the significant effect of inelastic scattering in thick specimens will now be considered. This approximation is referred to as the 'weak phase–weak amplitude' approximation, and it should be applied to images that show significant deviations from the weak phase approximation. These deviations may show up in the reconstructed images based on the simpler approximation as apparent structural variations between images taken at different defocus values.

4.6.1   The weak phase–weak amplitude approximation

The first reason for the breakdown of the weak phase approximation is that for thick specimens the expansion of $\exp(-i\eta)$ only as far as $1 - i\eta$ is invalid, because the phase shift η is not appreciably less than unity. Secondly, the phase object approximation does not naturally include the effect of inelastic scattering in causing an attenuation of the elastic wave. The effect of inelastic scattering on the elastic wave function is described by multiplying the object wave function $\exp(-i\eta)$ by $\exp(-\varepsilon)$ (Eq. (3.14); § 3.2):

$$\psi_0(r) = \exp\left[-i\eta(r)\right] \exp\left[-\varepsilon(r)\right] \tag{4.30}$$

where $\exp(-\varepsilon)$ represents the attenuation of the amplitude of the elastic wave by inelastic scattering. The weak phase–weak amplitude approximation corresponds to the expansion of Eq. (4.30) to first order terms in η and ε:

$$\psi_0(r) = 1 - i\eta(r) - \varepsilon(r) \qquad (4.31)$$

The thickness limits for which the expansion is valid for η have been discussed (§ 3.2.1); the thickness limit for neglecting terms in $\varepsilon^2/2$ in comparison to ε is much less restrictive than the limit for η and this limit may be as large as 50 nm for both stained and unstained specimens (Misell 1976). The weak phase–weak amplitude approximation can be applied to thicker specimens than the weak phase approximation alone because in the expansion of Eq. (4.30) the real term ε can be combined with the real second order term in η, $\eta^2/2$, as $-(\varepsilon + \eta^2/2)$. Then, provided that the specimen is thin enough to neglect other second order terms in η and ε, such as $\eta\varepsilon$, the thickness limit for the expansion of Eq. (4.30) may be extended to 30 nm for unstained specimens and to 10 nm for stained specimens. However, the ε in Eq. (4.31) cannot be interpreted directly in terms of the amplitude attenuation of ψ_0 because a significant contribution from $\eta^2/2$ has been included. In the following analysis ε represents the sum of ε and $\eta^2/2$. The diffracted wave $S(v)$ is calculated from the Fourier transform of Eq. (4.31):

$$S(v) = \delta(v) - iA(v) - E(v) \qquad (4.32)$$

and the intensity of the diffracted wave is ($v \neq 0$):

$$I(v) = |S(v)|^2 = |A(v)|^2 + |E(v)|^2 \qquad (4.33)$$

The image wave function $\psi_i(r)$ is calculated by multiplying $S(v)$ by the transfer function $T(v)$ and taking the inverse transform. The image intensity $j(r)$ is linearly related to $\eta(r)$ and $\varepsilon(r)$ in bright-field microscopy (Erickson and Klug 1971):

$$j(r) = 1 - 2\eta(r) * q(r) - 2\varepsilon(r) * q'(r) \qquad (4.34)$$

provided that non-linear (squared) terms in η and ε are neglected. The resolution functions q and q' are, respectively, the Fourier transforms of the transfer functions $\sin \gamma$ and $\cos \gamma$. The transform of the image intensity is given by:

$$J(v) = \delta(v) - 2A(v) \sin \gamma(v) - 2E(v) \cos \gamma(v) \qquad (4.35)$$

From a single image transform it is not possible to determine both A and E. In order to reconstruct both η and ε a second image, at least, is required recorded at a different defocus, thus altering the transfer functions $\sin \gamma$ and $\cos \gamma$ in Eq. (4.35). For two images j_1 and j_2 recorded at defocus values Δf_1 and Δf_2, respectively, the image transforms are:

$$\left.\begin{array}{l} J_1 = \delta - 2A \sin \gamma_1 - 2E \cos \gamma_1 \\ J_2 = \delta - 2A \sin \gamma_2 - 2E \cos \gamma_2 \end{array}\right\} \tag{4.36}$$

In principle these simultaneous equations can be solved for A and E. As in all situations involving the recording of more than one micrograph, it is assumed that the specimen structure (defined by η and ε) does not change between images. This is a severe limitation since radiation damage will certainly affect the high resolution components of $\eta(r)$; however, for crystalline specimens, images may be recorded from different areas of the specimen which have not previously been irradiated. From Eq. (4.36) the transforms A and E can be calculated (Frank 1972), subject to the limitation that A and E are indeterminate at zeros of the function $\sin (\gamma_2 - \gamma_1) = \sin (\pi \Delta f \lambda v^2)$, where $\Delta f = \Delta f_2 - \Delta f_1$. Thus in practice it is necessary to use at least three or four images when applying the weak phase–weak amplitude approximation to bright-field images (Erickson and Klug 1971; Frank 1972, 1973).

Electron diffraction data are of limited use here, because although the amplitude of the diffraction pattern $(|A|^2 + |E|^2)^{1/2}$ can be determined from Eq. (4.33), the phase angle of $S(v)$ cannot be simply determined from the image transforms, Eq. (4.36), as is possible for the weak phase approximation (§ 4.5).

When using the weak phase–weak amplitude approximation, the parameters C_s and Δf of the respective transfer functions must be determined from the image of the carbon support film. The carbon film must be thin (5–10 nm) so that its optical transform corresponds to that of a weak phase object; a thick carbon film may give significant amplitude contributions depending on $\cos \gamma$ (see Eq. 4.35) and the positions of the zero values of $\sin \gamma$ will be displaced, giving serious errors in C_s and Δf. The defocus cannot be determined to an accuracy better than the thickness t of the specimen, because the image transform will include a focus difference t between the top and bottom of the specimen (see § 7.3). Because of the dependence of $\gamma(v)$ on $\frac{1}{4}C_s v^4$ and $\frac{1}{2}\Delta f v^2$, Eq. (3.19), the accuracy with which C_s and Δf are determined must be very good at high spatial frequencies. Thus although a defocus error of 50 nm will give an error of only 0.15 rad in $\gamma(v)$ at 0.5 nm^{-1} (2 nm resolution), this error will be 0.58 rad at 1 nm^{-1} (1 nm resolution) and 2.32 rad at 2 nm^{-1} (0.5 nm resolution). Such large phase errors in $\sin \gamma$ and $\cos \gamma$ give correspondingly large errors in the phases of A and E calculated from the image transforms. For a phase error of less than 0.2 rad in $\gamma(v)$, Δf should be determined to within ± 17 nm at 1 nm resolution and to within ± 4 nm at 0.5 nm resolution.

Figure 4.27 shows three micrographs from a focus series of a negatively-stained catalase crystal together with their optical diffraction patterns (Erickson and Klug 1971). The right-hand half of the optical transform shows the diffraction pattern from the catalase whilst the left-hand side shows the transfer function used to determine the appropriate $\sin \gamma$ and $\cos \gamma$ functions. Note that since the highest order diffraction spot does not extend beyond about 0.5 nm^{-1} (2 nm resolution) the effect of the spherical aberration term on the value of γ is negligible and C_s need not be determined accurately. From the numerical transforms of these three images, values for A and E can be calculated using sets of equations such as Eq. (4.36). The reconstructed images corresponding to Figs. 4.27a, b and c, corrected for the effects of the transfer function, are shown in Figs. 4.28a, b and c as contour plots. The differences between these reconstructions, showing essentially the structure of $\eta(r)$, represent the limitations of the weak phase–weak amplitude approximation, but further analysis of the image transform data shows that the deviations from the weak phase approximation are even larger (Erickson and Klug 1971).

It is possible to obtain estimates for the relative magnitudes of the phase A and amplitude E contrast terms by taking a focus series of a crystalline specimen in an attempt to find a micrograph where the amplitude and phase contrast terms in Eq. (4.35) cancel for a particular reciprocal lattice point, that is:

$$A(v) \sin [\gamma(v)] + E(v) \cos [\gamma(v)] = 0 \qquad (4.37)$$

This usually occurs for overfocus so that the signs of $\sin \gamma$ and $\cos \gamma$ are opposite. For the unstained ribosome crystal data shown in Fig. 3.21 (§ 3.4.3), the (0, 1) diffraction order, corresponding to $v = 1/59.5$ nm^{-1}, is of negligible intensity for $\Delta f = -25 \, \mu m$ (overfocus). For such small spatial frequencies the effect of C_s may be neglected and $\gamma(v)$ calculated from:

$$\gamma = \frac{2\pi}{\lambda} \left(\frac{-\Delta f v^2 \lambda^2}{2} \right) = -\pi \Delta f v^2 \lambda$$

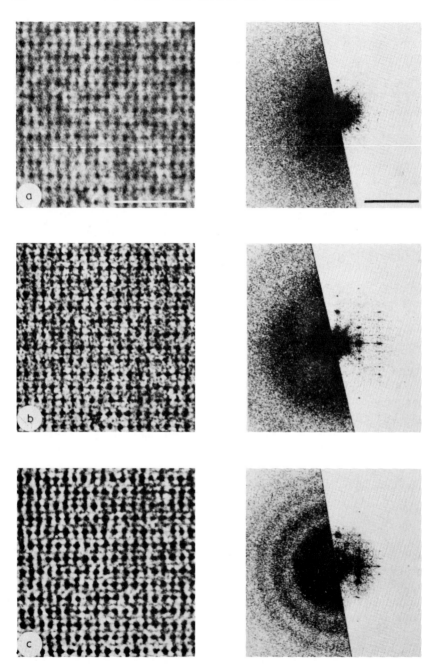

Fig. 4.27. Bright-field images of negatively-stained catalase and the corresponding optical diffraction patterns at different defocus: (a) 80 nm, (b) 540 nm, (c) 1450 nm underfocus. Image bar = 50 nm, diffraction bar = 0.4 nm^{-1}. (From Erickson and Klug 1971.)

Fig. 4.28. Noise filtered, average images reconstructed from computer calculated Fourier transforms of catalase images, corrected for lens aberrations and defocus; (a) 80 nm, (b) 540 nm, (c) 1450 nm underfocus. Image bar =.5 nm. (From Erickson and Klug 1971.)

using $\lambda = 3.7$ pm $(E_0 = 100 \text{ keV})$. Thus in Eq. (4.37), $\gamma = 0.08$ rad, so that the ratio $E/A \simeq 0.08$. Thus the amplitude term contributes approximately 8% to the image transform for this particular diffraction order. Such an investigation will indicate whether the weak phase or weak phase–weak amplitude approximation should be used in the analysis of an image.

4.7   Averaging and spatial filtering for non-periodic specimens

The purpose of image processing is to produce an increase in image contrast rather than an increase in resolution, although high resolution detail, previously masked by noise, may be revealed as a result of processing. Often image processing produces a loss in resolution as a result of averaging dissimilar or misaligned structural units and fine detail is lost in the averaging process. This is particularly relevant when considering ways of increasing the contrast and visibility of the structure of isolated molecules, such as virus subunits and enzymes. Whereas it is usually possible to obtain an objective measure of the resolution of an image of a translationally ordered specimen from its optical diffraction pattern or a rotationally symmetric specimen from its numerical image transform, there is no objective measure for isolated molecules. The transform of an isolated molecule will be continuous and, as such, indistinguishable from the transform of the carbon support film. There will be no effective cut-off of the image transform as for ordered specimens. So, although the microscope performance or resolution can be assessed, little can be said about the structural integrity of the isolated molecules. There is always the danger of over-interpreting fine detail; often an upper limit can be placed on the finest detail interpretable so that, for example, in negatively-stained specimens structural detail smaller than 2 nm should not be considered interpretable. Then spatial filtering can be performed on the image, excluding all information at spatial frequencies greater than $0.5 \, nm^{-1}$ from the reconstruction. Images of isolated molecules can always be corrected for the effect of defocus and lens aberrations either optically or numerically as described in § 4.4. This procedure does not depend on having an ordered specimen (Frank 1972), but it is much more difficult to combine images taken at different defocus to avoid frequency gaps in the transfer function, because the image transforms have to be aligned correctly. This alignment is relatively easy with two-dimensionally ordered specimens because the image transforms consist of spots on a well defined reciprocal lattice. The two main alternatives for increasing the contrast of isolated molecules are averaging; that is, adding together similar and aligned shapes either photographically (§ 4.7.1) or numerically (§ 4.7.2), or spatial filtering of images using the optical or numerical transform (§ 4.7.3).

4.7.1   Photographic superposition

The first problem encountered in averaging isolated molecules is finding similar molecules; unless the molecular geometry favours a particular orientation on the specimen grid, molecules will occur in all possible orientations. To select molecules in a similar orientation involves classifying shapes into particular groups. Each of these shapes represents a particular two-dimensional projection of a single three-dimensional structure, with the one reservation that in the case of negatively-stained specimens it is the two-dimensional stain distribution that is observed; this may not be a true representation of the actual shape of the three-dimensional molecule. The problem of selecting molecules from a whole field is shown in Fig. 4.29 where

---

Fig. 4.29.   Bright-field image of a field of isolated negatively-stained hexons from adenovirus. Image bar = 50 nm. Selected profiles were classified and printed at a larger magnification (image bar = 10 nm). Photographic superposition from the original negative is shown on the right-hand side. Note that the superpositions were made at an overall magnification of 2×10^6 for ease of alignment. (E.B. Brown and N.G. Wrigley, unpublished.)

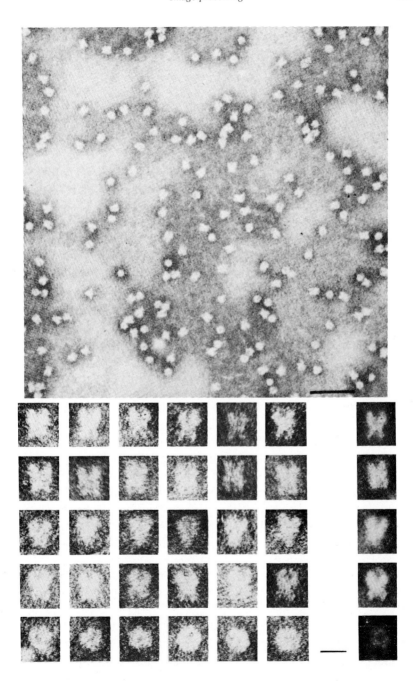

isolated hexon subunits from adenovirus (dimensions approximately 8 nm diameter by 12 nm length) are shown. The micrograph is often taken at too low a magnification to identify shapes and a photographic enlargement to an overall magnification of 1–2,000,000 is recommended. Molecules of similar shape are chosen from this enlargement; for the hexon it was possible to identify five characteristic shapes, as shown in Fig. 4.29. Other shapes, usually the majority, did not fit into any particular class. Of course, identifying only five projections is insufficient to produce a very high resolution, three-dimensional model of the hexon; for $N = 5$ in $N = \pi D/d$ with $D \simeq 8$ nm, the resolution of a three-dimensional reconstruction would be only about 5 nm. The number of projections required for a better resolution reconstruction may be reduced by a factor of 3 using the axial 3-fold symmetry of the hexon. However, in this particular example a three-dimensional reconstruction is not possible because the relative orientations of these five views are not known, without performing experiments in which the specimen is tilted.

Having identified similar projections of the molecule, they are added by photographic superposition as in photographic rotational averaging (§ 4.1.1). First a photographic enlargement ($\times 4$–10) of the original micrograph is made and molecules are identified by placing a characteristic mark with a felt-tip pen on the non-emulsion side of the plate. One particular molecule is projected onto card (overall magnification about 1–2,000,000, giving an image several centimetres in size) and the outline of the molecule is drawn onto the card. A sheet of photographic paper or film is placed under the card in a fixed position. For n superpositions, an exposure of this one particular molecule is made for $1/n$th of the time required to make a normal single exposure. A second molecule of a similar type to the first is brought to the centre by moving the film or plate in the enlarger; the card + photographic paper is moved until this next image falls within the outline drawn and a second exposure is made. In this way n similar molecules can be superimposed; the upper limit is about 5–10 superpositions, because of the difficulty of drawing accurately the original template on the card. The final superpositions for the hexon molecules are shown on the right-hand side of Fig. 4.29.

Photographic superposition is considerably easier if the molecule has some symmetry, such as the example shown in Fig. 4.30 for 2,3,4,5-tetra-acetoxymerithiophene which has four mercury atoms at the corners of a pentagon with a sulphur atom at the fifth corner (Ottensmeyer et al. 1973). The images shown in Fig. 4.30 (a–h) where selected from dark-field micro-

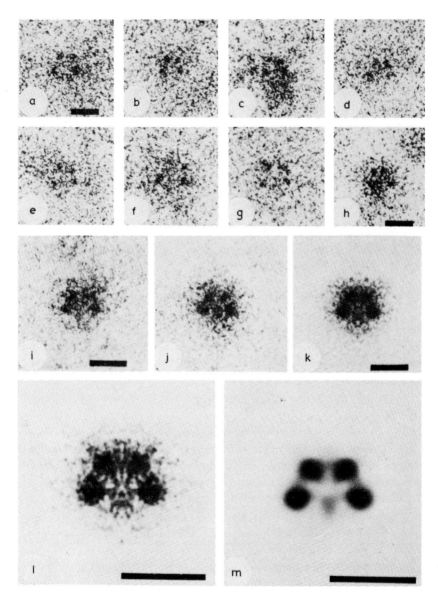

Fig. 4.30.   (a)–(h) Dark-field (tilt) images of eight individual molecules of 2,3,4,5 tetraacetoxy-merithiophene. (i)–(k) Visually aligned photographic superposition of (i) 4, (j) 16 and (k) 64 images ($k = 32$ plus 32 reflections). (l) Computer-assisted superposition of 64 images. (m) is (l) optically filtered to a resolution of 0.26 nm (low-pass filtering). Image bars = 1 nm. (From Ottensmeyer et al. 1973.)

graphs as being the most likely shapes of the molecule. Care had to be taken in the selection of molecules from such a noisy image; in this example the chosen shapes were influenced by the *known* structure of the organo-metallic compound. Of course, if something is known about the stereochemistry of the structure the information should be used to select views for averaging. Otherwise careful controls should be made, because selecting similar shapes that appeal to the eye can give interesting but completely false structures. As the number of superpositions increases from Fig. 4.30 (i–k) evidence of a fifth atom at the fifth corner of the pentagon, not observed in any of the individual images, appears. The equivalent computer superposition (Fig. 4.30 l), however, did show a lot of false detail. The final step carried out by Ottensmeyer et al. (1973) to reduce the noise still present in Fig. 4.30 l was to perform a spatial filtering on the optical transform of this image by excluding all information at spatial frequencies greater than $3.8 \, \text{nm}^{-1}$ (0.26 nm resolution), a low-pass filter (see § 4.7.3). The result shown in Fig. 4.30m is a fairly clear representation of four electron dense centres (the mercury atoms) and the fifth region of high density may be the relatively light sulphur atom. As an isolated occurrence the reader may be rather sceptical about the result but Ottensmeyer and co-workers (e.g. Ottensmeyer et al. 1972, 1975, 1977) have used similar and related techniques to produce images of molecular resolution for small molecules that are consistent with the stereochemistry of the molecules examined.

4.7.2   Computer superposition

The alignment of images by computer is more objective than visual methods because a quantitative measure of the accuracy of alignment and the similarity of aligned molecules can be obtained (Frank 1972, 1973). The procedure of aligning molecules both translationally and rotationally is called *cross-correlation*. Unfortunately the translational and rotational alignments have to be performed independently, and for illustration the procedure for translational alignment will be given.

As a prerequisite for computer superposition small image areas around each molecule are scanned to produce a set of digitised images. For example, with a molecule of maximum dimension 12 nm scanned at the equivalent of 0.5 nm steps a 32×32 matrix will suffice giving a reasonable margin of error for aligning the molecule on the densitometer; if it is difficult to see the molecule a 64×64 image scan should be made. Before two digitised images can be aligned for addition or subtraction (as for the image subtraction procedure in § 6.2), the image areas must be scaled so that the optical densities of both images represent the same total number of electrons detected on the photograph, otherwise different molecules will not be given equal weighting in the superposition. Thus for two molecules with image intensities $j_1(x, y)$ and $j_2(x, y)$ (32×32, 64×64 etc.) the summed densities of the scanned image must be such that:

$$\sum_x \sum_y j_1(x, y) = \sum_x \sum_y j_2(x, y) \tag{4.38}$$

This normalisation is achieved in the computer by the separate addition of the density values for each image, followed by multiplying one set of image intensities, say j_2, by the ratio of $\sum j_1 / \sum j_2$; this procedure can be repeated for all n images j_n by multiplying each image set by $\sum j_1 / \sum j_n$. In principle this type of scaling could be implemented in photographic superposition, but this is not easy – several photographic copies of each image must be made and the mean density levels evaluated using a microdensitometer.

To illustrate the properties of the *cross-correlation* function $c(x)$, the translational alignment of two image profiles $f(x)$ and $g(x)$ in one dimension is described. The cross-correlation function $c(x)$ is defined by:

$$c(x) = \int_{-\infty}^{+\infty} f(x_0)g(x + x_0)\,dx_0 \tag{4.39}$$

In practice the limits of the integral (summation) are taken over the extent of the images. The shape of $c(x)$ with respect to its peak position and the width of the curve give an indication of how good the correlation (alignment and similarity) of two images is, and indicates how far one image has to be moved to give the best alignment.

To see how this works in practice, first consider two profiles $f(x)$ and $g(x)$ that are identical and aligned (Fig. 4.31a). When $x = 0$ in Eq. (4.39), $c(0)$ is the sum of all the products $f(x_0)g(x_0)$ for all x_0 values; here the maxima and minima of both f and g coincide, and both positive and negative variations about the mean optical density (the x-axis) add to give a large contribution to $c(0)$. If g is now displaced by a small distance x, the maxima and minima will be slightly out of alignment, and some of the prominent features will be subtracted to give zero in the product $f(x_0)g(x + x_0)$; the sum $c(x)$ will therefore be smaller than $c(0)$. Increasing x correspondingly puts f and g

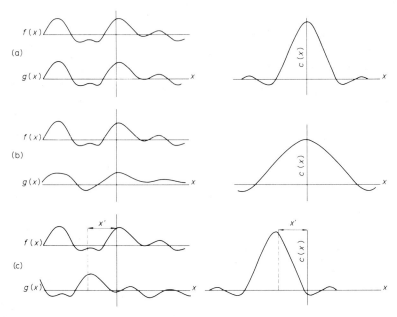

Fig. 4.31.   The cross-correlation $c(x)$ of two image profiles $f(x)$ and $g(x)$ when f and g are: (a) the same and aligned; (b) similar and aligned; (c) the same and misaligned by x'.

further out of register, and generally as x (the displacement of g with respect to f) increases, $c(x)$ will decrease. Thus $c(x)$ gives a measure of the similarity of two image profiles.

g is now made slightly different from f, or a low resolution version of f used as in Fig. 4.31b, but with the two images still aligned. Even with $x = 0$ in Eq. (4.39) the maxima and minima do not exactly coincide, and $c(0)$ is smaller than for the example shown in Fig. 4.31a. Also displacing g by a small distance x makes little difference to the sum of the products $f(x_0)g(x + x_0)$ because of the original differences that exist between f and g. However, as the displacement x increases, $c(x)$ does decrease because the two profiles become increasingly out of register. So the width and height of $c(x)$ show the similarity between the two profiles; a very broad $c(x)$ indicates two dissimilar images. If $g(x)$ is out of alignment with $f(x)$ by a distance x', as in Fig. 4.31c, with f and g now identical, then for $x = 0$ the sum of the products $f(x_0)g(x_0)$ will not be very large because maxima and minima will not coincide. However, as g is displaced to the right, the two curves approach alignment, and $c(x)$ will increase until $x = -x'$, when f and g are exactly aligned. The maximum in $c(x)$ then occurs at $x = -x'$, the extent of the original misalignment of the two images. The position of the maximum in the cross-correlation function $c(x)$, gives the translational misalignment of the two images, which can then be compensated for numerically by reinterpolating g onto a new coordinate system with its origin at $-x'$. The two images can then be added numerically.

The accuracy with which the peak position of $c(x)$ can be determined depends on the width of $c(x)$; if $c(x)$ is broad, the alignment (choice of x') cannot be made accurately. In practice, noise due to, for example, the carbon support film structure or intrinsic differences in the stain pattern around the molecules, will make cross-correlation less sensitive to image misalignment.

This analysis can be extended to two dimensions by calculating the two-dimensional cross-correlation function $c(x, y)$ of two images $f(x, y)$ and $g(x, y)$:

$$c(x, y) = \int_{-\infty}^{+\infty} \int_{-\infty}^{+\infty} f(x_0, y_0) g(x + x_0, y + y_0) \, dx_0 \, dy_0 \qquad (4.40)$$

when the maximum in $c(x, y)$ at (x', y') will give the image translational misalignment, which can be corrected for by interpolating the image array $g(x, y)$.

In practice the cross-correlation function for translational misalignment is calculated using the fast Fourier transform (FFT, see § 3.4.2) because the integrals in Eq. (4.40) become simple products of the Fourier transforms of f and g, F and G, respectively, with a significant reduction in computing times. The Fourier transform of Eq. (4.40) gives:

$$C(v_x, v_y) = F^+(v_x, v_y) G(v_x, v_y) \qquad (4.41)$$

where F^+ is the complex conjugate of F. Thus the transform of the two images f and g are calculated; the transform F^+ is calculated from F (if $F = a + ib$, $F^+ = a - ib$). The inverse transform of the product $F^+ G$ gives $c(x, y)$. However, if in addition to translational (lateral) misalignment, the images are rotationally misaligned, $c(x, y)$ will not have a well defined maximum and the translational alignment cannot be made successfully. The rotational alignment must first be made; this is done by calculating the moduli $|F|$ and $|G|$ of the two image transforms which are translationally but *not* rotationally invariant. The transform intensities $|F|^2$ and $|G|^2$ are rotationally correlated, transforming to polar coordinates, as in rotational filtering (§ 4.1). The angular rotation ϕ of $|G|^2$ with respect to $|F|^2$ necessary to produce a maximum in the power spectrum can be determined; ϕ corresponds to the original rotational misalignment of the images g and f. Thus g is interpolated in polar coordinates onto a new set of coordinates rotated by ϕ with respect to the original axes. The images g and f can now be translationally aligned.

Evidently visual methods of alignment of two images are much simpler because the eye judges both rotational and translational alignments simultaneously. For computer correlation it is desirable to align the images before scanning on the densitometer, so that the scans of the molecules are approximately rotationally aligned.

The procedure of translational and rotational correlation is repeated for all images to be superimposed; for each pair of images the width of the cross-correlation function will give an indication of the accuracy of alignment and the similarity of the images. In principle it should be possible to perform shape classification of molecules by such a procedure, but in practice the eye is best at making a preliminary selection of similar molecules. This information can, however, be quantitatively assessed using the numerical cross-correlation function.

4.7.3   Spatial filtering using the Fourier transform of the image

The image transform of an aperiodic specimen can be filtered so that only low resolution Fourier components are allowed to contribute to the reconstructed image. This elimination of high-frequency Fourier components, called *low-pass* spatial filtering, will remove from the image all fine detail specified by an upper frequency limit v_{max}. The important decision that has to be made is the demarcation between genuine structure (signal) and noise; noise due to the carbon support film and the photographic emulsion corresponds to high-frequency components of the image transform. However, other types of noise, such as the granular appearance of the negative stain, or the background scattering caused by the embedding medium in sectioned material, may not be clearly separated in the transform from the transform of the structure. It is usually impossible to make a clear division between signal and noise; reducing the frequency cut-off v_{max} may result in the loss of genuine fine structure. The choice of v_{max} for spatial filtering is made on the basis of the preservation of structure expected from a given specimen preparation procedure. For example v_{max} is unlikely to be greater than 0.5 nm^{-1} for sectioned material.

Spatial filtering can be done optically or by computer. For optical filtering, a mask with a single circular hole is made; its size is determined by measuring the diameter of the photographed optical diffraction pattern corresponding to v_{max}. In the numerical transform all transform values of (n, m) such that $(M \sqrt{(n^2 + m^2)})/(DN) > v_{max}$ are set to zero; for an image of $N \times N$ sample points and a densitometer scan spot size of $D(\mu m)$. Figure 4.32b shows the

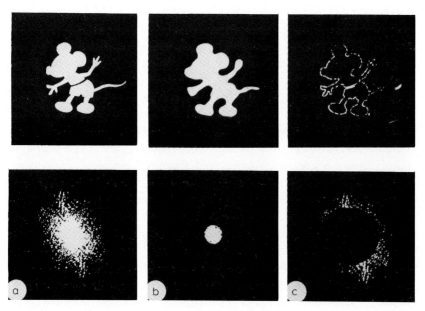

Fig. 4.32. Optical transforms: spatial filtering of 'Mickey Mouse': (a) original, (b) low-pass filtration, (c) dark-field edge enhancement. (From Harburn, Taylor and Welberry, Atlas of Optical Transforms, Bell and Hyman Ltd. 1975.)

effect of low-pass spatial filtering on a photograph of 'Mickey Mouse' (Fig. 4.32a); the lower part of Fig. 4.32 shows the original extent of the transform and the masked transform. In this optical simulation (Harburn et al. 1975) there is no noise on the image, so it is possible to see clearly what effect the exclusion of higher spatial frequencies has had on the appearance of the original object; the fine detail in the hands, for example, has been lost. This type of spatial filtering has been applied to dark-field images of organo-metallic compounds by Ottensmeyer et al. (1972, 1973). An example is shown in Fig. 4.30 l and m, where it can be seen that most of the false high resolution detail has been eliminated by spatial filtering.

It is also possible to produce filtered images where the low resolution information is excluded from the image transform in an attempt to see low contrast fine detail masked by a low resolution background structure; an application of this *high-pass* spatial filtering is the enhancement of the wall structure in bacterial cells, where the wall may not be easily visible and is difficult to measure. Allowing only high frequency components to contribute to the reconstructed image emphasizes fine detail and the results obtained from biological specimens should always be treated with caution. Always verify that the information is consistent with the structure seen in the original image. The exclusion of the zero-order and low order Fourier components produces results similar to dark-field microscopy, where the unscattered beam is excluded from the image (§ 5.4). High-pass spatial filtering emphasizes regions of the image where the optical density is changing most rapidly and it approximates to a differentiation of the original image. Figure 4.32c illustrates the result of excluding a large region of the Fourier spectrum of 'Mickey Mouse'; only regions of the image where the optical density is changing are observed. The same is true of the specimen shown in Fig. 4.33a, a scratched mica sheet (Harburn et al. 1975); the pattern of scratches is most clearly observed in the spatially filtered reconstruction Fig. 4.33b. At present there are no examples of the application of high-pass spatial filtering to electron micrographs, but it is a technique that is worth trying if suspected fine detail is masked by a large background contribution. It may be useful in delineating a molecule in negative stain.

Optically, high-pass filtered images are produced by making an opaque obstacle of the appropriate size to exclude all spatial frequencies less than v_{max}. However, the removal of the zero-order reference beam may result in a reconstruction that displays frequency doubling or half-spacings. The technique is also very sensitive to the presence of small particles of dust etc. on the micrograph. Numerical filtering is effected by setting all transform points (n, m) corresponding to $(M \sqrt{(n^2 + m^2)})/(DN) < v_{max}$ to zero and taking an inverse transform of this modified transform. Numerically it is also desirable to exclude very high frequency Fourier coefficients that arise from the carbon support film; so that, for example, the *band-limit* for filtering an image of negatively-stained molecules may extend from 0.5–1 nm^{-1}.

Ottensmeyer et al. (1977) have published an interesting procedure for filtering images of isolated molecules that appears to avoid the problems of separating the specimen transform from the noise. Nominally identical molecules are selected from the micrograph and photographic copies of these selected molecules are made onto film at a suitable magnification so that the molecular size is about 5 mm. These images are then cut out in

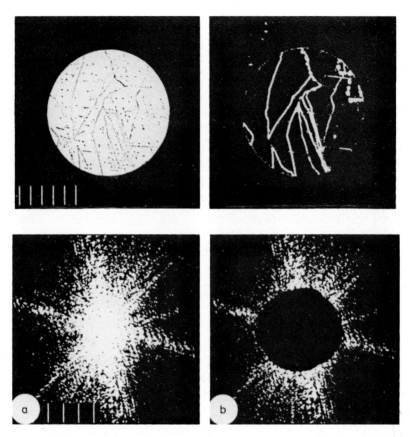

Fig. 4.33. Optical transforms: spatial filtering of scratches on mica: (a) original, (b) dark-field edge enhancement. (From Harburn, Taylor and Welberry, Atlas of Optical Transforms, Bell and Hyman Ltd. 1975.)

10 mm squares and mounted in an array so that they are aligned on a two-dimensional lattice as shown in Fig. 4.34a for 16 vasopressin molecules (Ottensmeyer et al. 1977); it is a considerable achievment to make the selection and alignment from such noisy images. The optical transform from such an array is obtained from a × 10 reduction in size of the original and it shows only 3 or 4 clear diffraction orders. A mask is then made with very small holes extending to 10–15 diffraction orders, although these high orders cannot be observed in the original transform. Thus in the reconstruction shown in Fig. 4.34b there are contributions from high resolution components of the Fourier spectrum of the original image array that produce the fine

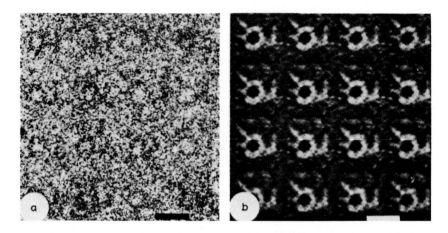

Fig. 4.34.   (a) Dark-field (tilt) image of an array of 16 individual iodinated vasopressin molecules. The molecule, which is about 1.7 nm in size, has nine amino acids, six of which form a ring. The iodine is attached to the tyrosine on the ring: (b) the reconstruction shows the ring of the molecule; the hooked tail of the three amino acids on the right and the iodinated side group of tyrosine are all visible. Image bar = 2 nm. (From Ottensmeyer et al. 1977.)

detail on the ring (6 amino acids) and tail (3 amino acids) of the vasopressin molecule. A reconstruction made from only the observed diffraction maxima produces only a rather blurred version of the vasopressin molecule; if the structure produced is genuine it must be concluded that the high frequency components, although indistinguishable from the noise in the original transform (optical or numerical), are present in the original image. The fact that the transform of the reconstructed image shows 10–15 distinct diffraction orders cannot be taken as evidence of the production of genuine high resolution detail; the nature of the filtering process with a mask comprising 10–15 diffraction orders is such that this result will always occur. In this example the structure is known, so that it is possible to verify that the fine detail is structurally correct. For example, the large spike on the ring is consistent with an iodine atom labelling the tyrosine; the rigid tail is consistent with the presence of the amino acid proline, and other fine structure could be due to the amino acid side chains. This sort of technique is limited to small, possibly planar, molecules which lie in a well defined three-dimensional orientation on the specimen grid support film; then the selection of 16 or more nominally identical molecules is possible. The result shown in Fig. 4.34b can be further refined using this reconstruction as a template for the original image and refining the orientations of the original vasopressin images. Because of the small mask holes used in the filtering procedure averaging occurs over the whole field (16 molecules), so that in the reconstruction (Fig. 4.34b) the molecules are identical; the maximum theoretical enhancement of the signal-to-noise ratio is the square root of the number of molecules in the array, $\sqrt{16} = 4$ in the example shown.

This procedure can also be implemented numerically by scanning the image array and producing a numerical transform; although only 2 or 3 diffraction orders are visible above the background, a reciprocal lattice can be chosen with the correct periodicity. The positions of higher order coefficients in the numerical transform can then be determined, although it is impossible to distinguish them from the background. A mask is numerically generated out to 10–15 diffraction orders with a mask size of 1×1 sample points to effect an averaging of the whole image. Unfortunately, because of the limited extent of the lattice, the numerical transform does not show very sharp maxima. This can be improved by optimum sampling of the image, either by matching the unit cell size to the size of the densitometer spot by a suitable choice of photographic reduction of the original image or by interpolation so that there is an exact number of sample points across a unit cell, and the image array is *boxed* numerically so that there are exactly 4×4 (say) repeat units in the digitised image. This latter procedure of boxing the image involves reducing the original digitised image area so that the background surrounding the periodic array is removed; the image then appears in the discrete Fourier transform as an exactly repeating structure. With these precautions the computer filtered result is of comparable quality to the optical result shown in Fig. 4.34.

References

Aebi, U., P.R. Smith, J. Dubochet, C. Henry and E. Kellenberger (1973), A study of the structure of the T-layer of *Bacillus brevis*, J. Supramol. Struct. *1*, 498–522.

Aebi, U., R.K.L. Bijlenga, B. ten Heggeler, J. Kistler, A.C. Steven and P.R. Smith (1976), A comparison of the structural and chemical composition of giant T-even phage heads, J. Supramol. Struct. *5*, 475–495.

Aebi, U., R. van Driel, R.K.L. Bijlenga, B. ten Heggeler, R. van den Broek, A.C. Steven and P.R. Smith (1977a), Capsid fine structure of T-even bacteriophage: binding and location of two dispensible capsid proteins into the P23* surface lattice, J. Mol. Biol. *110*, 687–698.

Aebi, U., B. ten Heggeler, L. Onorato, J. Kistler and M.K. Showe (1977b), New method for localizing proteins in periodic structures: Fab fragment labelling combined with image processing of electron micrographs. Proc. Natl. Acad. Sci. USA *74*, 5514–5518.

Blaurock, A.E. (1975), Bacteriorhodopsin: a trans-membrane pump containing α-helix, J. Mol. Biol. *93*, 139–158.

Burge, R.E. and R.F. Scott (1975), Binary filters for high resolution electron microscopy. I, Optik *43*, 53–64.

Burge, R.E. and R.F. Scott (1976), Binary filters for high resolution electron microscopy. II, Optik *44*, 159–172.

Chanzy, H., J.-M. Franc and D. Herbage (1976), High-angle electron diffraction of frozen hydrated collagen, Biochem. J. *153*, 139–140.

Crowther, R.A. (1976), The interpretation of images reconstructed from electron micrographs of biological particles, in: Proceedings of the Third John Innes Symposium, 'Structure-function relationships' of proteins', R. Markham and R.W. Horne, eds. (North-Holland, Amsterdam), pp. 15–25.

Crowther, R.A. and L.A. Amos (1971), Harmonic analysis of electron microscope images with rotational symmetry, J. Mol. Biol. *60*, 123–130.

Crowther, R.A. and R.M. Franklin (1972), The structure of the group of nine hexons from adenovirus, J. Mol. Biol. *68*, 181–184.

Crowther, R.A. and U.B. Sleytr (1977), An analysis of the fine structure of the surface layers from two strains of *Clostridia*, including correction for distorted images, J. Ultrastruct. Res. *58*, 41–49.

Crowther, R.A., E.V. Lenk, Y. Kikuchi and J. King (1977), Molecular reorganization in the hexagon to star transition of the baseplate of bacteriophage T4, J. Mol. Biol. *116*, 489–523.

DeRosier, D.J. and A. Klug (1968), Reconstruction of three dimensional structures from electron micrographs, Nature *217*, 130–134.

Erickson, H.P. (1974), Microtubule surface lattice and subunit structure and observations on reassembly, J. Cell. Biol. *60*, 153–167.

Erickson, H.P. and A. Klug (1971), Measurement and compensation of defocusing and aberrations by Fourier processing of electron micrographs, Phil. Trans. Roy. Soc. London B *261*, 105–118.

Frank, J. (1972), A study of heavy/light atom discrimination in bright-field electron microscopy using a computer, Biophys. J. *12*, 484–511.

Frank, J. (1973), Computer processing of electron micrographs, in: Advanced techniques in biological electron microscopy, J.K. Koehler, ed. (Springer-Verlag, Berlin), pp. 215–274.

Fraser, R.D.B., T.P. MacRae, E. Suzuki and C.L. Davey (1976), Image processing of electron micrographs of deformed filaments, J. Microsc. (Oxford) *108*, 343–348.

Glaeser, R.M. and L.W. Hobbs (1975), Radiation damage in stained catalase at low temperature, J. Microsc. (Oxford) *103*, 209–214.

Hahn, M.H. (1972), Eine optische Ortsfrequenzfilter- und Korrelationsanlage für electronen-mikroskopische Aufnahmen, Optik *35*, 326–337.

Hahn, M.H. and W. Baumeister (1973), Möglichkeiten und Grenzen electronenmikroskopischer Abbildung einzelner Atome in sub- und supramolekularen Systemen, Cytobiologie *7*, 224–243.

Harburn, G., C.A. Taylor and T.R. Welberry (1975), Atlas of optical transforms (Bell and Hyman Ltd., London).

Henderson, R. (1975), The structure of the purple membrane from *Halobacterium halobium*: Analysis of the X-ray diffraction pattern, J. Mol. Biol. *93*, 123–138.

Horne, R.W. and R. Markham (1972), Applications of optical diffraction and image reconstruction techniques to electron micrographs, in: Practical methods in electron microscopy, Vol. 1, part 2, A.M. Glauert, ed. (North-Holland, Amsterdam), pp. 327–434.

Hui, S.W., D.F. Parsons and M. Cowden (1974), Electron diffraction of wet phospholipid bilayers, Proc. Natl. Acad. Sci. USA *71*, 5068–5072.

Ishii, T. and M. Yanagida (1975), Molecular organization of the shell of the T_{even} bacteriophage head, J. Mol. Biol. *97*, 655–660.

Johansen, B.V. (1976), Bright-field electron microscopy of biological specimens. VI. Signal-to-noise ratio in specimens prepared on amorphous carbon and graphite crystal supports, Micron *7*, 157–170.

Kistler, J., U. Aebi and E. Kellenberger (1977), Freeze-drying and shadowing a two-dimensional periodic specimen. J. Ultrastruct. Res. *59*, 76–86.

Klug, A. and D.J. DeRosier (1966), Optical filtering of electron micrographs: reconstruction of one-sided images, Nature *212*, 29–32.

Kuo, I.A.M. and R.M. Glaeser (1975), Development of the methodology for low exposure, high resolution electron microscopy of biological specimens, Ultramicroscopy *1*, 53–66.

Lake, J.A. and K.R. Leonard (1974), Structure and protein distribution for the capsid of *Caulobacter crescentus* bacteriophage φCbK, J. Mol. Biol. *86*, 499–518.

Mellema, J.E. and H.J.N. van den Berg (1974), The quaternary structure of alfalfa mosaic virus, J. Supramol. Struct. *2*, 17–31.

Misell, D.L. (1976), On the validity of the weak-phase and other approximations in the analysis of electron microscope images, J. Phys. D: Appl. Phys. *9*, 1849–1866.

Ottensmeyer, F.P., E.E. Schmidt, T. Jack and J. Powell (1972), Molecular architecture: The optical treatment of dark field electron micrographs of atoms, J. Ultrastruct. Res. *40*, 546–555.

Ottensmeyer, F.P., E.E. Schmidt and A.J. Olbrecht (1973), Image of a sulfur atom, Science *179*, 175–176.

Ottensmeyer, F.P., R.F. Whiting, E.E. Schmidt and R.S. Clemens (1975), Electron micro-tephroscopy of proteins: A close look at the ashes of myokinase and protamine, J. Ultrastruct. Res. *52*, 193–201.

Ottensmeyer, F.P., J.W. Andrew, D.P. Bazett-Jones, A.S.K. Chan and J. Hewitt (1977), Signal-to-noise enhancement in dark-field electron micrographs of vasopressin: filtering of arrays of images in reciprocal space, J. Microsc. (Oxford) *109*, 259–268.

Smith, P.R. and J. Kistler (1977), Surface reliefs computed from micrographs of heavy metal shadowed specimens, J. Ultrastruct. Res. *61*, 124–133.

Steven, A.C., B. ten Heggeler, R. Müller, J. Kistler and J.P. Rosenbruch (1977), The ultra-structure of a periodic protein layer in the outer membrane of *Escherichia coli*, J. Cell Biol. *72*, 292–301.

Stroke, G.W. (1969), Image deblurring and aperture synthesis using *a posteriori* processing by Fourier-transform holography, Optica Acta *16*, 401–422.

Stroke, G.W. and M. Halioua (1972), Attainment of diffraction-limited imaging in high-resolution electron microscopy by '*a posteriori*' holographic image sharpening, I. Optik *35*, 50–65.

Stroke, G.W., M. Halioua, F. Thon and D. Willasch (1974), Image improvement in high-resolution electron microscopy using holographic image deconvolution, Optik *41*, 319–343.

Taylor, C.A. and J.K. Ranniko (1974), Problems in the use of selective optical spatial filtering to obtain enhanced information from electron micrographs, J. Microsc. (Oxford) *100*, 307–314.

Taylor, K.A. and R.M. Glaeser (1976), Electron microscopy of frozen hydrated biological specimens, J. Ultrastruct. Res. *55*, 448–456.

Unwin, P.N.T. and R. Henderson (1975), Molecular structure determination by electron microscopy of unstained crystalline specimens, J. Mol. Biol. *94*, 425–440.

Wurtz, M., J. Kistler and T. Hohn (1976), Surface structure of *in vitro* assembled bacteriophage lambda polyheads, J. Mol. Biol. *101*, 39–56.

Chapter 5

Instrumental methods
of contrast enhancement

The normal transmission electron microscope image of a biological specimen consists of a relatively high resolution elastic component formed by the elastically scattered electrons superimposed on a background, which arises from the inelastically scattered electrons, and is therefore blurred by chromatic aberration (§ 2.5.2). This inelastic contribution to the image causes not only a loss in image contrast but also a loss in image resolution, since high resolution detail of low contrast may ·be masked. These detrimental effects of inelastic scattering are more evident for unstained biological macromolecules than for heavy atom labelled macromolecules, because the inelastic scattering decreases relative to the elastic scattering as the atomic number of the specimen increases. Because of the nature of inelastic electron scattering, the inelastic image contains structural information of low resolution (2 nm); its effect can be reduced significantly for ordered specimens by using the Fourier filtering techniques described in Chapter 4. There are two instrumental developments that will successfully separate the·inelastic and elastic components of the image. The energy-selecting electron microscope (§ 5.2), which is a normal electron microscope fitted with an electron spectrometer below the objective lens or first projector lens (Henkelman and Ottensmeyer 1974; Egerton et al. 1975), discriminates between elastically and inelastically scattered electrons on the basis of their energy differences. This discrimination can be made irrespective of the thickness of the specimen. Secondly, in the scanning transmission electron microscope (STEM) perfected by Crewe and his co-workers (Crewe and Wall 1970; Crewe 1971) (§ 5.3) all lenses precede the specimen, there are none after it, so the inelastic image is free from chromatic aberration. The principle of separation of the elastic and inelastic signals is based on the difference in the angular distributions of the two scattering processes (§ 2.5.2). Essentially the inelastic scattering occurs at small angles (< 1 mrad), whereas elastic scattering predominates at larger angles of scattering (> 10 mrad). An annular detector with a hole corresponding to a maximum angle of about 10 mrad allows the inelastic and unscattered electrons (which are undeviated in passing through the specimen) to be detected separately from the elastic signal, which is collected on the annulus. The unscattered and inelastic components can be separated by an electron spectrometer beneath the detector. This discrimination on the basis of angular differences in elastic and inelastic scattering is not valid for thick specimens, where multiple scattering may result in inelastic scattering at relatively large angles.

Whether the image contrast arises from phase contrast or scattering contrast depends on the defocus value, Δf, in relation to the specimen spacing r. Phase contrast requires the objective lens to be defocused; an approximate defocus $\Delta f = r^2/2\lambda$ can be used to enhance spacings of r for electrons of wavelength λ (§ 3.2.2). Thus for specimens exhibiting low resolution of only 5 nm, the defocus Δf required is 3.4 μm, and is well away from the near-focus conditions normally used in the electron microscopy of thin biological sections. Near-focus images of low resolution specimens show only very low phase contrast because the phase contrast transfer function $\sin \gamma$ is virtually zero over the range of spatial frequencies exhibited by normal sections or intact cells dried down on a grid. Consequently, provided that the image of a low resolution specimen is taken just under focus (0–200 nm), the main contribution to the image contrast will arise from *scattering* or *aperture* contrast, which depends on the differential scattering of electrons from regions of the specimen with differing mass thickness (mass per unit area of specimen) ρt (§ 5.1). Scattering contrast depends on the effective size of the objective aperture, and this can be used to enhance the differential scattering effects (Agar et al. 1974); a small objective aperture, semi-angle α subtended at the specimen, intercepts relatively more electrons scattered from regions of the specimen with high density. One way of reducing the value of α is to increase the focal length of the objective lens by increasing the length of the specimen holder for a top-entry specimen stage (Lovell and Chapman 1975). This modification causes a deterioration in image resolution, but this is usually not important for most biological sections (Misell 1976).

Although dark-field microscopy is commonly used in metallurgy and materials science, it is not often employed in biological electron microscopy. One of the disadvantages of bright-field microscopy is the lack of contrast in the image as a result of a large contribution to the background from the unscattered beam. In dark-field microscopy this unscattered component is intercepted by the objective aperture (Dupouy 1967; Johnson and Parsons 1969; Kleinschmidt 1971; Dubochet 1973); the inelastically scattered component, which has very little effect on bright-field image contrast, is the main reason for limiting the gain in contrast obtainable in dark-field microscopy (§ 5.4). There are certain disadvantages of dark-field microscopy that offset this advantage of increased contrast, including image interpretation problems, which arise because the image intensity is not linearly related to the specimen structure, and the problem of the increase in electron dose to the specimen necessary to obtain the same number of electrons in the image as the corresponding bright-field image; this latter problem is related to distinguishing the structure (signal) from the noise in the image (§ 5.4.1).

An electron-optical technique similar to *Schlieren microscopy* (Goodman 1968) can be used to enhance contrast in a particular direction across the specimen. The electron illumination is tilted so that the unscattered beam just passes inside the objective aperture, but one half of the diffraction pattern is intercepted by the aperture (§ 5.6). Because of the similarity of the images to those of heavy metal-shadowed specimens, in which surface features are enhanced, this technique has been referred to as *optical shadowing* (Haydon and Lemons 1972) or *topographical imaging* (Cullis and Maher 1974, 1975). More technically, the terms *single-sideband holography* or *half-plane imaging* have been used to describe the fact that only a single side (half-plane) of the diffraction pattern is used for imaging (Hanszen 1973). Whilst this technique is of undoubted value for contrast enhancement, images formed from only one half of the diffraction pattern must be interpreted with caution.

One of the problems with images in which contrast is a result of phase contrast is the difficulty of choosing a single defocus value Δf that will give a $\sin \gamma$ that is uniform over a large range of spatial frequencies. Numerically this problem can be solved by combining image transforms from a focus series (§ 4.4 to § 4.6). Experimentally, this *extended* phase contrast can be achieved by superimposing two images *in situ* taken at two different defocus values (Johansen 1975). The first image is recorded with a defocus value which maximises the phase contrast function $\sin \gamma$ at low spatial frequencies and then a second image, taken at optimum defocus ($\simeq 100$ nm underfocus) in which $\sin \gamma$ extends to high spatial frequencies, is superimposed on the first exposure. This technique gives an image with an effective $\sin \gamma$ that is uniform over a larger spatial frequency range than either of the constituent images (§ 5.7). Specimen drift and contamination must be very low for the success of this technique.

5.1   Increasing the focal length of the objective lens

Scattering contrast has sometimes been misleadingly referred to as absorption contrast giving the impression that image contrast arises from the absorption of electrons in the specimen. In fact for normal electron microscope specimens (10–100 nm thick), nearly all the incident electrons are transmitted by the specimen and it is the objective aperture that removes electrons from the scattered beam. The fraction of the incident beam intensity, I_0, that is stopped by the objective aperture depends on the semi-angle α mrad subtended by the objective aperture at the specimen (Fig. 5.1).

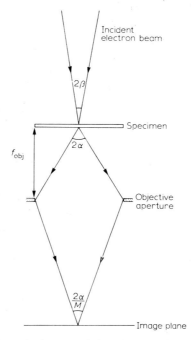

Fig. 5.1.   The various planes in the transmission electron microscope. β is the semi-angle subtended by the illumination at the specimen; α is the semi-angle subtended by the objective aperture at the specimen and f_{obj} is the focal length of the objective lens. (From Misell 1976.)

Decreasing α preferentially reduces the contribution to the image from regions of the specimen with high mass thickness, ρt, which scatter electrons into larger angles than regions of low mass thickness. All the unscattered electrons, which are undeviated, and most of the inelastically scattered electrons, pass through the objective aperture. Normally the angle of illumination β is much smaller than α. The ratio of the electron intensity scattered within the objective aperture $I(\alpha)$ to the incident electron beam intensity I_0 determines the contrast C in the image:

$$C = (I_0 - I(\alpha))/I_0 \qquad (5.1)$$

$I(\alpha)$ depends on the mass thickness ρt of the specimen and the *mass scattering cross-section*, S_p, which depends on the electron scattering properties of the specimen, the objective aperture semi-angle α and the incident electron beam energy E_0. Fortunately S_p is not too dependent on the atomic number Z of the scattering material, so that theoretical values of S_p calculated for one element can be used for both unstained and stained biological material. In Fig. 5.2 theoretical values for $\log_{10}(I(\alpha)/I_0)$ have been plotted against ρt for three values of α and $E_0 = 40\,\text{keV}$ (Misell and Burdett 1977). This linear relation establishes the following exponential relationship between $I(\alpha)/I_0$ and ρt:

$$I(\alpha)/I_0 = \exp(-S_p\rho t) \qquad (5.2)$$

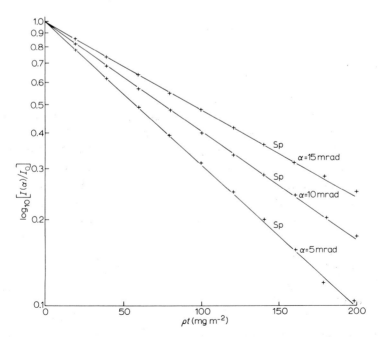

Fig. 5.2.   The linear variation of $\log_{10}(I(\alpha)/I_0)$ with mass thickness ρt (mg m^{-2}). The gradient of each line gives the constant S_p from $2.303\log_{10}(I/I_0) = -S_p\rho t$ for a given E_0 and α. ($+$) denotes theoretical values for $\log_{10}(I/I_0)$. (From Misell and Burdett 1977.)

S_p increases with decreasing α and decreasing E_0, and so scattering contrast is enhanced by the use of a small α and a low incident electron energy. $I(\alpha)/I_0$ is not linearly related to ρt, so that Eq. (5.2) provides a basis for determining density variations ρ in the specimen, since the thickness t of the specimen is usually uniform over the area imaged. Table 5.1 lists values of

TABLE 5.1

Theoretical mass scattering cross-section S_p (m^2 mg^{-1}) for scattering within an objective aperture of semi-angle α (mrad)

E_0 (keV)	$\alpha = 5$	$\alpha = 10$	$\alpha = 15$
20	0.0237	0.0192	0.0164
40	0.0116	0.0089	0.0072
60	0.0078	0.0058	0.0043
80	0.0061	0.0043	0.0033
100	0.0051	0.0035	0.0025

S_p is used as a parameter in determining mass thickness ρt from $I(\alpha)/I_0 = \exp(-S_p\rho t)$

S_p for a set of α and E_0 values. The units of S_p are m^2 mg^{-1} and the units of ρt are mg m^{-2} (10 mg m^{-2} = 1 μg cm^{-2}). The measurements necessary to determine the ratio $I(\alpha)/I_0$ from an image are shown in Fig. 5.3(a). Only the ratio $I(\alpha)/I_0$ is required and the absolute values of $I(\alpha)$ and I_0 do not have to be determined. The image is scanned in lines using a microdensitometer with a spot size that is significantly larger than any (high-resolution) phase contrast structure from the carbon film supporting the specimen. In considering contrast arising from scattering contrast it is implicitly assumed that there is no contribution to the image from phase contrast. $I(\alpha)$ and I_0 values are both measured with respect to the unexposed level on the photographic image (usually at the edge); I_0, representing the incident beam intensity, is measured at a hole in the carbon supporting film or on the supporting film, provided that the film is thin (10 nm), whilst $I(\alpha)$ is determined from the optical density of the image of the specimen region of interest. Figure 5.3b shows a composite scan across a micrograph of a stained section of *Bacillus subtilis* including regions of the stained section, the embedding medium and the carbon support film alone (Fig. 5.3a). In the example shown, $I(\alpha)/I_0$ in the region of the stained section was approximately 0.39 so that using the appropriate value for S_p of 0.0072 mg m^{-2} ($\alpha = 15$ mrad,

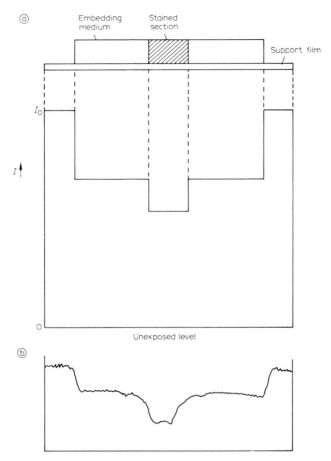

Fig. 5.3. (a) Schematic diagram showing the measurements made in order to determine the mass thickness of a stained section in an embedding medium. (b) Microdensitometer scan across a stained section of *Bacillus subtilis* corresponding approximately to (a). (From Misell and Burdett 1977.)

$E_0 = 40 \, \text{keV}$), $\rho t = 131 \, \text{mg m}^{-2}$. The density ρ cannot be determined from ρt unless the specimen thickness t is known; an estimated section thickness of 50 nm gives a value of $\rho = 2620 \, \text{kg m}^{-3}$ (2.62 g cm^{-3}). In addition to determining the variations in ρt across a biological section, the mass of dried biological material, such as intact bacteria, can be measured. The product of the average value of ρt and the cross-sectional area of the particle, A (corrected for the magnification of the micrograph), $\rho t A = $ density × volume of specimen = mass.

In order to maximise scattering contrast and increase sensitivity to small changes in ρt it is necessary to use as small a value of α (Fig. 5.1) as possible, corresponding to a large value of S_p in Eq. (5.2). There is a lower limit to α beyond which there is no significant increase in contrast. The maximum *theoretical* value for the improvement in relative image contrast from using a small value for α is about a factor of 2.4. α is determined geometrically from the diameter, d_{obj}, of the objective aperture and the objective focal length, f_{obj} (Fig. 5.1):

$$\alpha = d_{obj}/2f_{obj} \qquad (5.3)$$

so that α can be decreased either by decreasing d_{obj} or increasing f_{obj}. In a microscope with a side-entry specimen stage (Agar et al. 1974) the only possible way of decreasing α is by using a smaller objective aperture. There is, however, a lower limit to d_{obj} because of contamination and associated astigmatism problems that occur with an objective aperture much smaller than $20\,\mu m$; for an objective lens of focal length 1.6 mm this corresponds to $\alpha = 6.2$ mrad. Note that the actual value d_{obj} can be measured using an optical microscope, whilst f_{obj} is given in the manufacturer's brochure. Figure 5.4 shows the gain in contrast that can be obtained on stained sectioned material by decreasing α from (a) 17.5 mrad ($d_{obj} = 56\,\mu m$) to (d) 3.8 mrad ($d_{obj} = 12\,\mu m$); quantitatively this corresponds to a relative increase in contrast by a factor of 2 from Fig. 5.4a to Fig. 5.4d.

The alternative of increasing f_{obj} applies only to a top-entry specimen stage (Agar et al. 1974); by lengthening the tip of the specimen holder, f_{obj} can be increased from its normal value of about 2 mm to 10–15 mm. Thus it is possible to use a large objective aperture (say $70\,\mu m$) and still attain α values as small as 3 mrad (Lovell and Chapman 1975). There are two problems, however, in using a *high contrast* specimen holder: firstly, the normal range of the objective lens current may be insufficient to focus the image. This problem can usually be solved by using a given accelerating voltage setting E_0 (say 80 keV) whilst using an objective lens setting appropriate to a lower value of E_0 (say 60 keV). Secondly, the increase in f_{obj} causes proportional increases in the spherical and chromatic aberration constants, C_s and C_c, of the objective lens. In principle this should not cause an appreciable loss in image resolution because the maximum off-axis angle α is correspondingly decreased. The experimental results of Lovell and Chapman (1975) do show, however, a deterioration in the microscope resolution from 0.34 nm for $f_{obj} = 2$ mm to nearly 4 nm for $f_{obj} = 14$ mm.

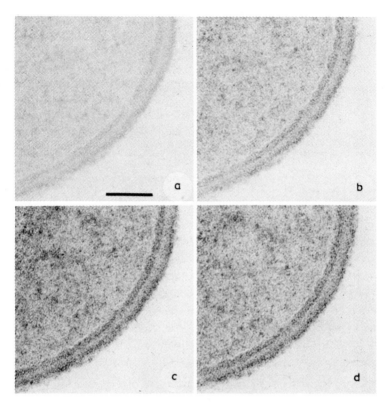

Fig. 5.4. Stained section of *Bacillus subtilis* showing the variation of image contrast with objective aperture size α: (a) $\alpha = 17.5$ mrad, (b) $\alpha = 8.5$ mrad, (c) $\alpha = 6.1$ mrad, (d) $\alpha = 3.8$ mrad. Image bar $= 50$ nm, $E_0 = 80$ keV. (E.B. Brown, unpublished.)

Scattering contrast may also be increased by lowering the incident electron energy E_0 because S_p increases with decreasing E_0. The relative increase in contrast in changing from $E_0 = 80$ keV to 40 keV is about 2. An example of the contrast gain in a stained section of *Bacillus subtilis*, obtained by decreasing E_0 is shown in Fig. 5.5 for (a) 80 keV, (b) 60 keV and (c) 40 keV (Misell and Burdett 1977). The main disadvantage in decreasing the incident electron energy is an increase in chromatic aberration, r_c, since:

$$r_c = C_c \Delta E \theta / E_0 \tag{5.4}$$

Thus decreasing E_0 by a factor of two increases r_c by a factor of two; the chromatic aberration constant C_c and the energy loss ΔE do not change

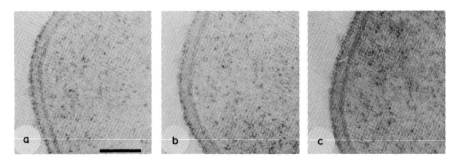

Fig. 5.5.   Stained section of *Bacillus subtilis* showing the variation of image contrast with incident electron energy, E_0: (a) $E_0 = 80\,\text{keV}$, (b) $E_0 = 60\,\text{keV}$, (c) $E_0 = 40\,\text{keV}$. Image bar $= 100\,\text{nm}$. (From Misell and Burdett 1977.)

significantly. In addition the average angle of scattering θ increases because the incident electron beam interacts more strongly with the specimen as E_0 decreases. This is one reason why the use of high accelerating voltages (80–100 keV) is recommended for the examination of normal sections, so that the image resolution is not limited by chromatic aberration. The loss in contrast resulting from using higher values of E_0 can be readily offset by using a smaller objective aperture than normal.

5.2   *Applications of the energy-selecting electron microscope*

The energy-selecting electron microscope can produce normal (unfiltered) and energy-filtered (elastic only) images by means of an electron spectrometer fitted below the objective or first projector (intermediate) lens of a conventional transmission electron microscope (Henkelman and Ottensmeyer 1974; Egerton et al. 1975). This instrument is not available commercially, because to some extent its function is duplicated by the STEM; the addition of an electron spectrometer to the electron optics of a normal electron microscope may degrade its spatial resolution, whereas in the STEM the electron spectrometer does not affect the optics of the microscope (§ 5.3). However, energy discrimination rather than the angular separation used in the STEM is the most satisfactory way of separating the elastic and inelastic components of the scattered electron beam.

The largest gain in contrast by using an energy-selecting electron microscope is with specimens composed of a large proportion of low atomic weight material, such as biological material; the ratio of inelastic scattering to elastic scattering in carbon ($Z = 6$) can be as large as 10:1 (Burge 1973).

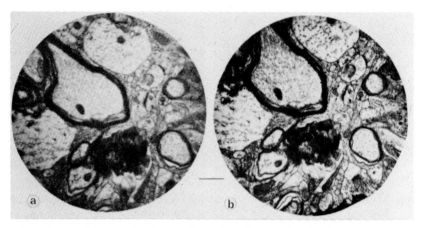

Fig. 5.6. Images of visual cortex tissue sections, about 60 nm thick, in bright-field microscopy: (a) normal (elastic + inelastic) image, (b) filtered (elastic only) image. Image bar = 1 μm. (R.F. Egerton, unpublished.)

Figure 5.6 shows (a) normal and (b) energy-filtered (elastic) images of a thick ($t = 60$ nm) stained section of visual cortex tissue taken in bright-field conditions (Egerton 1976). The gain in contrast is evident, although no additional fine detail is revealed in the filtered image, because of the low intrinsic resolution of the specimen. An even greater gain in contrast can be obtained from unstained biological specimens because the inelastic scattering represents a major contribution to the image. Figure 5.7 is (a) a normal and (b) an elastic image of polyoma virus, unstained and unshadowed, taken in dark-field conditions (Henkelman and Ottensmeyer 1974). In the normal image (Fig. 5.7a), the predominant inelastic scattering contributes to an overall blurring of the virus particle; in the elastic image (Fig. 5.7b), surface detail of about 3.5 nm resolution can be observed, although this detail is difficult to interpret because of the superposition of structure from the top and bottom of the virus.

5.3 Scanning transmission electron microscopy (STEM)

In the STEM a small electron probe (0.5 nm) is produced by a field emission gun and a condenser-objective lens (Crewe and Wall 1970; Crewe 1971). The electron probe is scanned across the specimen in a two-dimensional *raster* and the electron scattering from the specimen for each position of the scan spot is measured by a fixed detector (Fig. 5.8a); the image can be

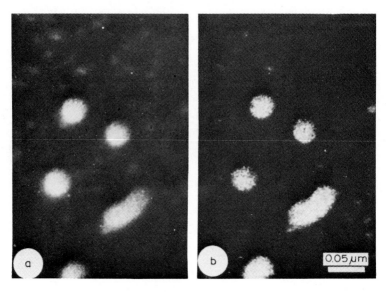

Fig. 5.7.   Images of polyoma virus, critical-point dried, unstained and unshadowed, in dark-field microscopy: (a) normal image, (b) filtered image. Image bar = 50 nm. (From Henkelman and Ottensmeyer 1974.)

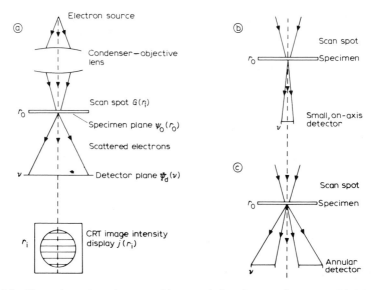

Fig. 5.8.   The various planes in a scanning transmission electron microscope with (a) a full detector, (b) a small on-axis detector and (c) an annular detector.

observed on a cathode ray tube (CRT) synchronised with the scan coils. There are no lenses after the specimen, so the inelastic scattering in the specimen is not affected by chromatic aberration, but because of the nature of the inelastic scattering process, the inelastic image is still of intrinsically low resolution (Isaacson et al. 1974). In normal use there is a small on-axis detector (Fig. 5.8b), the equivalent of bright-field microscopy or an annular detector (Fig. 5.8c), the equivalent of dark-field microscopy (§ 5.4). The annular detector allows the unscattered and inelastically scattered electrons to pass through the centre of the detector into an electron spectrometer, where the inelastic component can be separated from the unscattered component, and an inelastic image formed simultaneously with the elastic image collected on the annulus of the detector. There is also the possibility of forming simultaneous dark-field and bright-field images by using a small on-axis detector immediately below the annular detector. Note that since the angle of convergence of the incident electron beam is normally quite large (10 mrad), the semi-angle subtended by the centre hole of the annular detector at the specimen should be significantly larger than 10 mrad, so that all the unscattered and most of the inelastically scattered electrons pass through the hole. Collection of the unscattered electrons on the detector lowers image contrast because the elastic signal is relatively small (proportional to η^2; see § 5.3.1). In fact it can be seen that if a full detector is used, as in Fig. 5.8a, image contrast will be virtually zero because all electrons incident on the specimen will be detected in the transmitted signal for a fairly thin specimen (10–100 nm), irrespective of density variations in the specimen.

5.3.1 Theory of image formation in the STEM

Although it may be unnecessary to understand the theory of image formation in the STEM to use the instrument effectively, it is useful to derive a relationship between the image and the specimen for comparison with image formation in the conventional transmission electron microscope (CTEM).

In Fig. 5.8a, r_i defines the position of the scan spot in the specimen plane, with a resolution function $G(r_i)$ dependent on the aberrations and defocus, Δf, of the condenser-objective lens in the same way as the resolution function in the CTEM depends on $\gamma(v)$ (Thomson 1973). The specimen wavefunction at r_0, $\psi_0(r_0)$, describes the effect of the specimen on the incident electron beam; then the wavefunction, $\psi_i(r_i)$, immediately after the specimen at a point r_i will be (Misell et al. 1974):

$$\psi_i(r_i) = \psi_0(r_0)G(r_i - r_0) \tag{5.5}$$

In the STEM the electron scattering from the specimen is collected in the detector plane v. The electron wavefunction $\Psi_d(v)$ in the detector plane v is derived from the Fourier transform of ψ_i with respect to r_0, that is

$$\Psi_d(v) = \int \psi_0(r_0)G(r_i - r_0) \exp(2\pi i v \cdot r_0)\,dr_0 \qquad (5.6)$$

Equation (5.6) represents the angular distribution of the scattered electrons in the detector plane, including the broadening (convolution) effect of the convergent incident beam. The image intensity $j(r_i)$, corresponding to the scan spot position r_i, is calculated by integrating the intensity $|\Psi_d|^2$ over the field of the detector; this $j(r_i)$ is displayed sequentially on a CRT screen. Two detector geometries are considered for a weak phase object ($\psi_0 \simeq 1 - i\eta$). Using a small on-axis detector (Fig. 5.8b) only a small fraction of the total elastic scattering and larger fractions of the unscattered and inelastically scattered components are collected. Ignoring the inelastic component, this geometry in the STEM is equivalent to bright-field *coherent* image formation in the CTEM (§ 3.2) and the image intensity is given by:

$$j(r_i) = 1 - 2\eta(r_i) * q(r_i) \qquad (5.7)$$

where $q(r_i)$ is the Fourier transform of $\sin\gamma(v)$, defined by Eq. (3.19) (§ 3.2) for the condenser-objective lens. However, this detector geometry is seldom used in the STEM because of the low detection efficiency (about 10%) of a small on-axis detector as compared with a collector efficiency of about 90% in bright-field CTEM.

The normal detector geometry (Crewe 1971) consists of an annular detector (Fig. 5.8c), and the unscattered component and most of the inelastic component pass through the centre hole, while the annulus of the detector collects 80–90% of the total of the elastically scattered electrons. Thus integrating $|\Psi_d|^2$ over a large detector gives an image intensity distribution:

$$j(r_i) = \eta^2(r_i) * |G(r_i)|^2 \qquad (5.8)$$

which is a convolution of η^2 with the resolution function $|G|^2$ appropriate to the scan spot. This type of image is then equivalent to *incoherent* dark-field CTEM; this should be distinguished from normal dark-field microscopy in the CTEM, which corresponds to *coherent* imaging (see § 5.4). In incoherent imaging there are no interference effects and the image is free from phase contrast effects (Engel et al. 1974). Thus, although there is a non-linear relationship between the image intensity and η, it is not as complicated as the corresponding non-linear result in coherent dark-field microscopy (Eq. (3.25); § 3.2). In fact Eq. (5.8) which is a convolution integral can be solved using Fourier transforms to give η^2 (see Eq. (3.22); § 3.2), unlike Eq. (3.25) (Stroke and Halioua 1972). Note that Eq. (5.8) is an approximation, because account has not been taken of the hole in the annular detector in carrying out the integration of $|\Psi_d|^2$; this can lead to additional, but small, linear terms in η (Cowley 1975a). The detector efficiency in dark-field STEM is extremely high, because the annular detector can be large enough to collect virtually all the elastically scattered electrons from the specimen; this is in contrast to dark-field CTEM where the maximum angle of scattering for electrons that can be collected to form an image is limited by the objective lens aberrations. The contrast in dark-field STEM can be very high, because the only significant contribution to the background is a result of elastic scattering from the specimen support film; the inelastic scattering present in dark-field CTEM is virtually eliminated from the dark-field STEM image. Thus in a fair comparison of the relative merits of STEM and CTEM, the most efficient methods of imaging should be considered; that is, dark-field STEM and bright-field CTEM (Misell 1977). In these equivalent configurations the radiation dose given to the specimen in the two instruments is similar for a given signal-to-noise ratio in the image (see § 5.4.1), but the contrast in dark-field STEM may be significantly greater than bright-field CTEM, because of the relatively small background in the former system (Crewe et al. 1975). It should be noted that, whereas the signal in bright-field CTEM depends on 2η ($\simeq 0.2$), the signal in dark-field STEM is only η^2 ($\simeq 0.01$), so that it does not need much background to significantly reduce image contrast in the STEM. That is why it is important that none of the unscattered component and only a small fraction of the total inelastic scattered component is collected on the annulus. As an extreme example, collection of the whole of the unscattered component will give a signal proportional to η^2 ($\simeq 0.01$) on a background of unity; that is, an undetectable signal. In the applications described in the following section, it will be evident that the STEM is most effective in detecting single heavy atoms (§ 5.3.3), and only marginally better in examining unstained and negatively-stained biological specimens (§ 5.3.2 and § 5.3.3).

The STEM has one important practical advantage over the CTEM: focusing can be made objectively. The focus and stigmator controls of the objective lens are adjusted so as to obtain the smallest circular spot on the

CRT display or to quote Crewe 'to focus and correct astigmatism in the STEM merely involves making the image of a single heavy atom, viewed at several million magnification, as round and as sharp as possible; that is provided atoms are round!'

The greatest disadvantage of STEM for biological electron microscopy is that the information collected in a STEM picture is about 1/100th of that which can be obtained on the photographic plate used in the CTEM. Since only about one photographic plate in 10 yields a picture good enough for image analysis and processing, it may be necessary to photograph 1000 images in the STEM before a satisfactory picture is obtained.

5.3.2   *Applications of STEM*

In Fig. 5.9 a comparison of dark-field STEM images (a and b) of negatively-stained tobacco mosaic virus (TMV) stacked disc protein is made with a bright-field CTEM image (c) (Engel et al. 1976); the latter is printed with reversed contrast for the comparison, so that 'protein' (the stain exclusion regions) corresponds to dark areas on the print. One of the most evident differences between the STEM and CTEM images is the absence of phase contrast effects in the background surrounding the TMV particle in the STEM images. The visual differences between Fig. 5.9a and c (both minimal radiation dose images) are not so evident in their respective optical diffraction patterns (Fig. 5.9d and e), where the layer line structures from the helical particle are similar. Again the importance of recording images, even of negatively-stained specimens, under minimal radiation conditions is emphasized by the differences between Fig. 5.9a ($n_0 = 1500\,\mathrm{e}^-\,\mathrm{nm}^{-2}$ at the specimen) and Fig. 5.9b ($n_0 = 16{,}500\,\mathrm{e}^-\,\mathrm{nm}^{-2}$).

Figure 5.10 is an example of simultaneously recorded dark- and bright-field images of a negatively-stained giant T4 phage tail (Tschopp and Smith 1977) recorded using an annular detector with a separate on-axis detector immediately below the hole in the annulus. The images in Fig. 5.10 were obtained after processing the original images by Fourier filtering using a *low-pass* filter (§ 4.7.3) with an effective frequency cut-off at $1\,\mathrm{nm}^{-1}$ (1 nm resolution).

For negatively-stained specimens there seems to be little gain in using the STEM, except that images can be stored in digital form and processed directly by an *on-line* computer, without the necessity of first scanning a photographic image on a densitometer and then proceeding with *off-line* computer processing. Similar considerations apply to normal biological

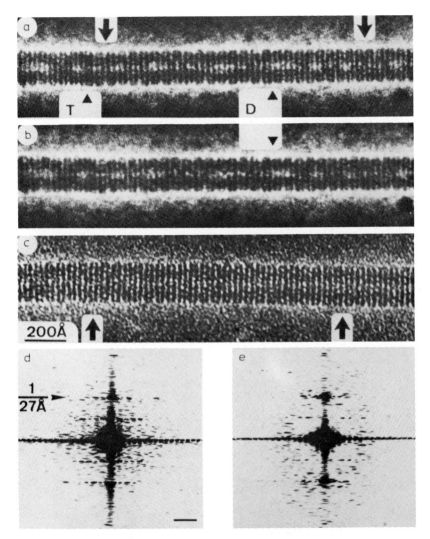

Fig. 5.9. Stacked disc aggregates of TMV protein negatively-stained with uranyl acetate: (a) recorded at the first scan in dark-field STEM (pre-irradiation dose $= 100\,e^-\,nm^{-2}$, recording dose $= 1500\,e^-\,nm^{-2}$); (b) recorded at the twelfth scan (previous dose, therefore $= 16,500$ $e^-\,nm^{-2}$, recording dose $= 1500\,e^-\,nm^{-2}$); (c) recorded in bright-field CTEM under minimal beam conditions (total dose $= 1600\,e^-\,nm^{-2}$). To facilitate comparison with the STEM images the bright-field image is printed with reversed contrast. The region between the arrows has been used for optical diffraction. Note that the specimen is better preserved in (a) than in (b) – the deformation of the interface between discs at the point D can be clearly seen. Pairing of the rings is easily visible in (a). A triplet structure, where three rings are packed together is recognisable at T. (d) Optical transform of (a), (e) optical transform of (c). Image bar $= 20\,nm$, diffraction bar $= 0.2\,nm^{-1}$. (From Engel et al. 1976.)

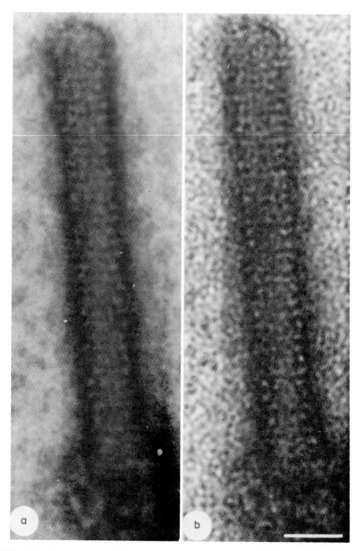

Fig. 5.10. Simultaneously recorded STEM dark-field (a) and bright-field (b) images of a uranyl acetate stained extra-long T4 tail. Radiation dose $= 30,000\,e^-\,nm^{-2}$, scan spot size $= 0.5\,nm$, scan step $= 0.4\,nm$; image bar $= 20\,nm$. (A. Engel and P.R. Smith, unpublished.)

sections with no clear advantage of STEM. However, images of thick biological sections (up to 1 μm thick), which would normally be seriously blurred by chromatic aberration in the CTEM, can be imaged in the STEM without any chromatic aberration, because of the absence of lenses after the specimen (Gentsch et al. 1974). Discrimination between the elastic and inelastic components on an angular basis is, however, not possible because the inelastic scattering spreads out to large angles. For thick specimens the resolution of STEM images is limited by *electron beam broadening* as a result of multiple electron scattering in the specimen (Cowley 1975b; Reimer and Gentsch 1975).

5.3.3 Separation of inelastic and elastic components

One of the most impressive applications of elastic/inelastic discrimination in the STEM is in resolving single heavy atoms. Figure 5.11 shows (a) an elastic

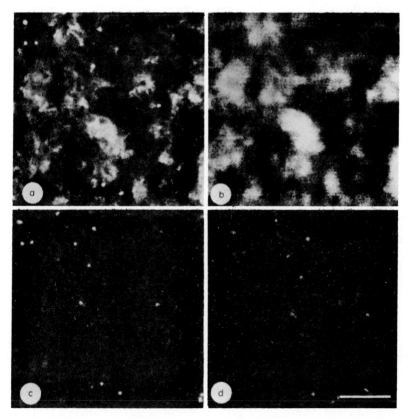

Fig. 5.11. STEM images of single mercury atoms on a thin carbon film; (a) elastic image, (b) inelastic image, (c) ratio (elastic/inelastic) image, (d) difference image. Image bar = 5 nm. (From Crewe et al. 1975.)

and (b) an inelastic image of mercury atoms on a carbon support film. Whereas the detail, particularly the atoms, is observable in the elastic image (Fig. 5.11a), the detail in the inelastic image (Fig. 5.11b) is blurred and heavy atoms cannot be distinguished from the background. Even Fig. 5.11a is a noisy image but it can be enhanced, either by taking an elastic/inelastic, (E/I), ratio image (Fig. 5.11c) or a difference image (Fig. 5.11d), $E-fI$, where f is an arbitrary fraction of the inelastic image intensity (Crewe 1971; Crewe et al. 1975). Both these types of enhancement assume that the noise in the image, arising mainly from the specimen support film, is similar in both the elastic and inelastic images; this assumption is validated by the similar appearance of the background structure in images Fig. 5.11a and b, although (b) looks like a blurred version of (a). In either example, the ratio or the difference image, the mercury atoms are clearly visible. The importance of the ratio image, I/E, is that its intensity is proportional to the atomic number Z of the particular atom (Crewe 1971). Thus Z-*number contrast* can be used to distinguish between different heavy atoms in a structure; it also enables a distinction to be made between contaminant atoms and atoms used, for example, to label specific functional groups in macromolecules, such as a nucleic acid.

Figure 5.12 is an example of (a) elastic and (b) inelastic images of unstained ribosomes from *Escherichia coli*; the contrast in both images is very good, but the inelastic image is not significantly inferior to the elastic image, and

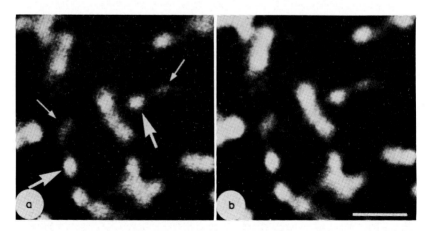

Fig. 5.12. (a) Elastic and (b) inelastic images of ribosomes from *Escherichia coli*, prepared in 10^{-3} M $MgCl_2$. The magnification is not high enough to see any detail and most of the molecules have aggregated. Small arrowheads indicate 30 S subunits; large arrowheads, 50 S subunits. Image bar = 50 nm. (J. Langmore, unpublished.)

neiter a ratio nor a difference image would lead to a significant improvement in the detail observed. This is an example where the resolution of the image is limited by the intrinsic resolution of the particular specimen; the fine detail in the relatively large ribosome structure may have been destroyed by the dehydration process and by radiation damage. This example confirms that for low resolution specimens (2 nm) the STEM, with its potential for producing elastic and inelastic images, will not produce significantly better images than the CTEM; for such specimens the inelastic image is not greatly inferior to the elastic image, which is intrinsically also of low resolution.

The potential for separating elastic and inelastic components in the STEM is likely to prove most useful in microanalysis, using the inelastically scattered electrons to form an *electron energy loss spectrum* (§ 2.5.2) characteristic of a very small area (1–10 nm^2) of the specimen (Isaacson and Johnson 1975).

5.4   Dark-field microscopy

In dark-field microscopy or *strioscopy* the unscattered component of the electron beam is intercepted by the objective aperture. Experimentally this may be achieved by tilting the incident electron beam, so that the unscattered beam falls outside the objective aperture, or a centre stop may be fixed across the objective aperture. Alternatively hollow-cone illumination, obtained by using an annular condenser aperture and an appropriately matching objective aperture, can be used, so that the cone of unscattered electrons falls outside the objective aperture. These different types of dark-field microscopy and their relative merits will be discussed in detail in § 5.4.2. The first decision to be made is whether dark-field microscopy has significant advantages over normal bright-field microscopy for examining biological specimens. The contrast should be higher for both unstained and stained biological specimens, because the unscattered background has been eliminated from the image. Against this, the inelastic scattering, which is relatively unimportant compared with the signal 2η in bright-field microscopy, is significant compared to the signal η^2 in dark-field microscopy. The other factor which is relevant is the electron radiation dose n_0 given to the specimen in order to obtain a given signal-to-noise ratio, S/N, in the image. For a given value of n_0 the number of electrons detected in dark-field is much smaller than in bright-field microscopy and the S/N value in dark-field will be much smaller. In order to achieve a comparable S/N ratio to bright-field, the radiation dose to the specimen may have to be increased by a factor of as much as 30 (§ 5.4.1). However, the results of Ottensmeyer and co-workers (Ottensmeyer et al.

1975a, 1975b, 1977) indicate that for small molecules, such as myokinase, protamine and vasopressin, the 'ashes' of the structure may retain their integrity rather like fossils! The most serious doubts about dark-field microscopy relate to image interpretation; the non-linear relation between the image intensity and structure η can cause artefacts in the image (Hanszen 1969), particularly with tilted illumination where the image is formed from only one half of the diffraction pattern. In this situation the objective aperture is asymmetrical with respect to the electron diffraction pattern of the specimen. In the study of amorphous materials using tilted illumination, fringes observed in the image may be incorrectly interpreted as evidence of small crystallites in the amorphous film. These 'lattice' fringes have been shown to be a result of the particular electron optics and not characteristic of the specimen (Krakow et al. 1976). However, these doubts about the validity of dark-field images are mitigated by the results of Ottensmeyer et al. on small (1–2 nm) structures, where for all the molecules, selected images were consistent with the known stereochemistry. It may be that this is a result of examining only small structures under virtually incoherent imaging conditions, so that the interference or phase contrast effects that can cause such artefacts (Hanszen 1969) are absent. Ultrastructure seen in dark-field microscopy should always be treated with caution, particularly if similar features cannot be observed in corresponding bright-field micrographs with a uniform transfer function $\sin \gamma$ (§ 3.2.2).

5.4.1 Dark-field versus bright-field microscopy

The interpretation problem arises in coherent dark-field microscopy partly because the image intensity depends on $(\eta - \eta_m)^2$, where η_m is the mean value of η over the area of specimen that is *coherently* illuminated. The extent of the specimen, d_{coh}, that is coherently illuminated depends on the angle of electron illumination β (Fig. 5.1), so that $d_{coh} \simeq 0.61\lambda/\beta$ for electrons of wavelength λ (Barnett 1974). Thus for well collimated illumination β will be 0.1 mrad and $d_{coh} = 23$ nm, for $\lambda = 3.7$ pm and $E_0 = 100$ keV. In this example electrons elastically scattered from regions of the specimen separated by distances as large as 23 nm may interfere and give phase contrast effects. However, if β is increased to 5 mrad, by using an extended electron source and a large condenser lens aperture, the coherence patch, d_{coh}, will only be 0.5 nm and detail in the specimen separated by distances larger than 0.5 nm will be *incoherently* imaged. Consequently the image will be free from phase contrast effects, as in dark-field STEM. The fact that the image intensity in coherent dark-field depends on $(\eta - \eta_m)^2$ means that it is impossible to distinguish between regions of low density (low η) or high density (high η) in the specimen, because the squared values of $(\eta - \eta_m)$ may be similar (Cowley 1975c). There is no such problem in bright-field microscopy, because the image intensity is proportional to $(\eta - \eta_m)$, and it is always possible to distinguish between high density $(\eta > \eta_m)$ and low density $(\eta < \eta_m)$ regions of the specimen. In incoherent imaging conditions, where each specimen point scatters independently, the definition of a mean value of η_m taken over a large area of specimen is not meaningful. Under these conditions the image intensity is proportional to η^2 at each point in the image. Also in-focus or near-focus dark-field images of low resolution specimens should not cause any interpretation problems, because the image can be considered to be formed by the differential scattering of electrons from high and low density regions of the specimen (see § 5.1). In order to give quantitative information on the contrast, C, and signal-to-noise ratio, S/N, in bright-field and dark-field images, it is necessary to calculate the contributions from the elastic scattering by the specimen (signal), and from the inelastic scattering by the specimen, and both elastic and inelastic scattering by the specimen support film (noise). Statistical noise, which is proportional to the square root of the number of electrons detected in the image, results from the random arrival of electrons in the image (sometimes referred to as 'electron noise'). It is inappropriate to give the details of such calculations here, since it is the final results that are relevant. However, the following information will give the physical basis of such calculations.

In bright-field microscopy, neglecting lens aberrations, the image contrast is 2η (background is unity), whilst the number of electrons detected is virtually n_0. Thus the signal S is $2\eta n_0$ and the signal-to-noise ratio in the absence of a support film is (Misell 1977):

$$S/N = 2\eta n_0/\sqrt{n_0} = 2\eta\sqrt{n_0} \tag{5.9}$$

In dark-field microscopy the contrast is η^2 divided by a background, arising from inelastic scattering in the specimen (j_i) and scattering from the support film (j_s). The total number of electrons detected is $n_0(\eta^2 + j_i + j_s)$, so that the signal-to-noise ratio in the image is:

$$S/N = \frac{n_0\eta^2}{\sqrt{n_0(\eta^2 + j_i + j_s)}} = \frac{\eta^2\sqrt{n_0}}{\sqrt{(\eta^2 + j_i + j_s)}} \tag{5.10}$$

Since j_i for unstained specimens is significantly greater than η^2 in Eq. (5.10), S/N in dark-field microscopy is much smaller than the corresponding bright-field S/N value. In order to increase S/N by a factor of k the value of n_0 must be increased by a factor of k^2. Thus, to obtain comparable S/N values in dark- and bright-field microscopy, the radiation dose n_0 to the specimen must be increased substantially in dark-field. Table 5.2 summarises calcula-

TABLE 5.2

A comparison of the contrast C ($\%$) and signal-to-noise ratio, S/N, in bright-field and dark-field microscopy for an electron dose $n_0 = 2000\,\mathrm{e}^-\,\mathrm{nm}^{-2}$ ($6.2\,\mathrm{e}^-\,\mathrm{nm}^{-2} = 1\,\mathrm{Cm}^{-2}$)

(a) Specimen: unstained biological material ($Z = 6$) of thickness 5 nm

| Substrate thickness (nm) | Bright-field | | Dark-field | | | | | |
| | | | + Inelastic | | | 30% Inelastic | | |
	C	S/N	C	S/N	F	C	S/N	F
0	20%	8.6	15%	1.6	30	36%	2.5	11
1	20%	8.6	12%	1.4	35	30%	2.3	15
2	19%	8.5	10%	1.4	40	26%	2.1	16
5	19%	8.3	7%	1.1	55	18%	1.8	21
10	18%	8.2	5%	1.0	80	12%	1.4	34

(b) Specimen: heavy atoms or a small group of atoms ($Z = 79$) of thickness 0.5 nm

| Substrate thickness (nm) | Bright-field | | Dark-field | | | | | |
| | | | + Inelastic | | | 30% Inelastic | | |
	C	S/N	C	S/N	F	C	S/N	F
0	30%	13.0	65%	5.1	7	86%	5.9	5
1	30%	12.9	47%	4.2	9	71%	5.4	6
2	29%	12.7	36%	3.8	12	60%	4.8	7
5	28%	12.6	22%	2.9	19	42%	4.1	9
10	27%	12.3	14%	2.5	25	29%	3.4	13

Incident electron energy $E_0 = 100\,\mathrm{keV}$, objective aperture semi-angle $\alpha = 0.01$ rad. F is the factor in electron dose required to give the same S/N value in dark-field as in bright-field microscopy.

tions of the contrast, C, and S/N for an electron dose corresponding to minimum irradiation conditions ($n_0 = 2000\,\mathrm{e}^-\,\mathrm{nm}^{-2}$), and for two types of specimen: (a) an unstained biological specimen of thickness 5 nm; and (b) a group of heavy atoms of thickness 0.5 nm. In the dark-field examples, results are shown assuming, firstly, that all the inelastically scattered electrons contribute to the image and, secondly, that 70% of the small-angle inelastic scattering near to the unscattered beam is stopped by the objective aperture. Clearly this latter condition is necessary if the contrast C for an unstained specimen is to be significantly better than the bright-field contrast. In either case, in dark-field operation, the electron dose factor, F, necessary to give an acceptable signal-to-noise ratio (at least 5) in the dark-field microscopy of unstained specimens is ten or more times greater than in bright-field microscopy. Also, it can be noted that, because of the small signal (η^2), electron scattering from the specimen support film (j_s) should be kept to a minimum by using very thin (1–2 nm) support films.

The dark-field contrast for a specimen consisting of heavy atoms (Table 5.2b) is significantly better than bright-field contrast, provided that the support film thickness is less than 5 nm. This improvement in contrast in dark-field results from the increase in elastic scattering (proportional to $Z^{3/2}$), relative to the inelastic scattering (proportional to $Z^{1/2}$). The signal-to-noise values are acceptable, and the increase in electron dose necessary to obtain a signal-to-noise ratio comparable to bright-field is only about 5. Thus, when resolving the heavy atoms used to label macromolecules, dark-field microscopy has a considerable advantage in terms of image contrast, with or without the reduction in the inelastic background. However, the advantages of dark-field microscopy for the examination of unstained biological macromolecules are not evident, unless the specimen support film is very thin, and additionally some effort is made to reduce the inelastic contribution to the image (see also § 6.2). The contrast advantage of dark-field microscopy diminishes as the specimen thickness increases, because the contribution from the inelastic scattering increases relative to the elastic scattering. This increase in inelastic scattering reduces the dark-field image contrast, but does not significantly affect the contrast in bright-field images (Ottensmeyer and Pear 1975). The examples of dark-field images of sectioned biological material both stained and unstained (§ 5.4.3), and of other unstained biological specimens to be shown in later sections (§ 5.4.4 and § 5.4.5) illustrate the potential of dark-field electron microscopy.

It should be noted that in dark-field STEM the signal is still proportional to η^2 but the contribution to the noise from the inelastic scattering can be virtually reduced to zero, so the image contrast is very high and the signal-to-noise in the image is approximately $\eta^2 n_0 / \sqrt{(\eta^2 n_0)} = \eta \sqrt{n_0}$ which is only a factor of two lower than the bright-field image signal-to-noise ratio.

5.4.2   *Different types of dark-field microscopy*

The simplest method of obtaining a dark-field image is to tilt the incident illumination, so that the unscattered beam is intercepted by the objective aperture as shown in Fig. 5.13c. The angle of tilt required is only about

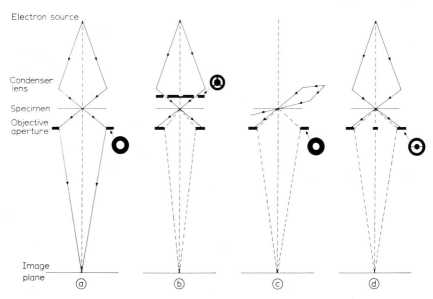

Fig. 5.13.   Schematic electron optics for bright- and dark-field microscopy: (a) conventional bright-field, (b) dark-field using an annular condenser aperture, (c) dark-field by tilting (the illumination angle is exaggerated), (d) axial dark-field – using an objective aperture with a centre stop. (From Kleinschmidt 1971.)

10 mrad (0.6°), and this small deflection of the incident beam is produced by a set of electromagnetic deflector coils; note that the tilt angle shown in Fig. 5.13c is exaggerated for illustrative purposes. The dark-field tilt controls can be adjusted, so that with the microscope in the diffraction mode, the main beam falls outside the objective aperture. Since the inelastic electron scattering predominates at small angles of scattering, a substantial fraction of the inelastic component will be removed from the image. The disadvantage of dark-field microscopy using tilted illumination is that the objective aperture is placed asymmetrically with respect to the electron diffraction pattern from the specimen; one-half of the elastic scattering is intercepted by the objective aperture. Because of this asymmetry, dark-field tilt images may appear to be astigmatic when the image has, in fact, been corrected for

astigmatism. An advantage of dark-field tilt, however, is the ease with which the microscopist can change from dark-field to bright-field microscopy by switching off the dark-field deflector system. Consequently the dark-field controls can be pre-set to their correct values and the microscopist can switch directly from bright-field to dark-field or *vice versa*. Alternating between bright-field and dark-field microscopy is not so easy with either a centre-stop objective aperture or an annular condenser aperture (see below). The alternative of mechanically displacing the objective aperture, so that the main beam is intercepted, is not recommended, because the off-axis aberrations of the objective lens (such as spherical aberration) will be very large and may result in poor images. An objective aperture with a centre stop can also be used to achieve dark-field imaging (Fig. 5.13d); a thin wire ($\simeq 5$–10 μm in diameter) is welded across the centre of the objective aperture (Dupouy 1967; Johnson and Parsons 1969). Such a centre stop intercepts electrons scattered at angles between 1–2 mrad, depending on the focal length of the objective lens; this centre stop may intercept as much as 70% of the inelastically scattered electrons. The use of this type of objective aperture for dark-field microscopy has the advantage that the symmetry of the diffraction pattern, with respect to the objective aperture, is retained. But the aperture has to be precisely aligned in the diffraction plane of the microscope; severe astigmatism can be caused by the electrical charging of the centre stop as a result of intercepting the intense unscattered beam. It is also very difficult to change directly from dark-field to bright-field microscopy by mechanically changing to a normal objective aperture, because the astigmatism and focus will both alter. In fact changing from bright-field to dark-field microscopy is usually impossible, because the mechanical alignment on most aperture controls is not sufficiently accurate.

Probably the ideal method for producing dark-field images is to use an annular (strioscopic) condenser aperture fitted in the second condenser aperture holder (Fig. 5.13b). The aperture has a centre stop of about 1–2 mm diameter with an annulus of about 200 μm width (Dupouy et al. 1969; Thon and Willasch 1972). Thus a cone of illumination subtending an angle of about 10 mrad at the specimen is produced. The objective aperture size is chosen so that the illumination cone just falls outside the aperture; scattered electrons then pass through the objective aperture, although as in all previous configurations, a significant proportion of the inelastically scattered electrons near to the illumination cone are intercepted by the objective aperture. Because the effective angle of illumination β is large, such dark-field images are formed under incoherent illumination conditions, and the images are free from interference or phase contrast effects. The objective aperture is placed symmetrically with respect to the diffraction pattern, so that the problems of asymmetry encountered in dark-field tilt are absent. It is relatively easy to change from dark-field to bright-field microscopy by changing from a strioscopic to a normal condenser aperture, but some adjustments to the condenser lens currents are required. The change from bright-field to dark-field cannot, however, be made quickly.

Focusing in dark-field microscopy is difficult. Unlike bright-field microscopy, the granular structure of a carbon film is not evident in dark-field, using either axial or tilted illumination. Also there is no objective test of the defocus level, since the optical transform of the image has no simple relation to the state of focus. Reference to Eq. (3.25) in § 3.2 for the image intensity j in dark-field shows that the image transform will not be simply related to either $\sin \gamma$ or $\cos \gamma$. Optical diffractograms of dark-field images show only a uniform disc almost independent of defocus, although astigmatism and specimen drift show up as distortions of the disc. In many biological applications the accuracy of achieving focus may not be critical, and it is sufficient to focus in bright-field, followed by the appropriate operation to give the dark-field image; however, this change in electron optics can give focus changes of 100–200 nm. Cowley et al. (1974) suggest that dark-field images can be focused by observing maximum contrast at a hole in a carbon film. At a magnification of $100,000 \times$ this maximum contrast position, corresponding to about 30 nm underfocus, can be determined with an accuracy of ± 20 nm; this is quite close to optimum defocus, about 50 nm underfocus, in axial or tilted dark-field microscopy.

5.4.3   Application of dark-field microscopy to sectioned material

A comparison of (a) bright-field and (b) dark-field (tilt) images of a stained liver section (Fig. 5.14) (Ottensmeyer and Pear 1975) shows that there is a significant gain in contrast in dark-field, but similar detail, such as the mitochondria, is visible in both images. However, for a thin 'unstained' section (OsO_4 post-fixed), the bright-field image (Fig. 5.14c) is poor and the dark-field image (Fig. 5.14d) is significantly better. It is difficult to compare completely unstained sectioned material, because without osmium post-fixation the specimen is poorly preserved. It is evident that for thin sections the gain in contrast in dark-field microscopy is significant enough to make it a useful technique for examining sections. Specimens which show low contrast in bright-field may in fact reveal valuable information on ultrastructure in

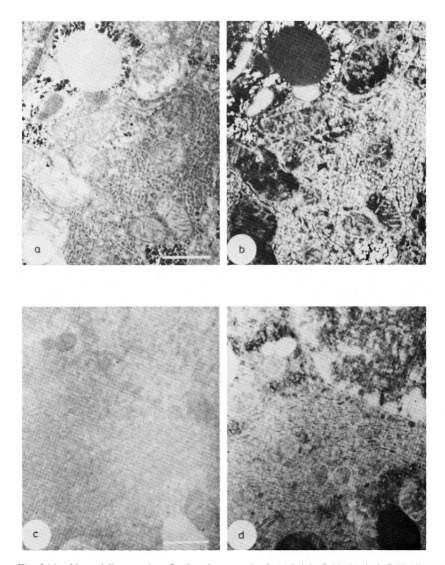

Fig. 5.14. Normal liver section, fixed and post-stained: (a) bright-field, (b) dark-field (tilt). Image bar $= 1 \ \mu$m, $E_0 = 60$ keV. Normal liver section, glutaraldehyde fixed, OsO_4 post-fixed, unstained: (c) bright-field, (d) dark-field (tilt). Image bar $= 1 \ \mu$m, $E_0 = 80$ keV. (From Otten-smeyer and Pear 1975.)

dark-field. The gain in contrast by using dark-field microscopy diminishes for thicker (50 nm) stained sections; for such specimens the bright-field image contrast is adequate, and the convenience of focusing and correcting astigmatism is to the advantage of bright-field microscopy.

5.4.4   Application of dark-field microscopy to unstained specimens

Dark-field microscopy is particularly recommended for examining unstained nucleic acids. For example, Fig. 5.15 shows a comparison of (a) bright-field

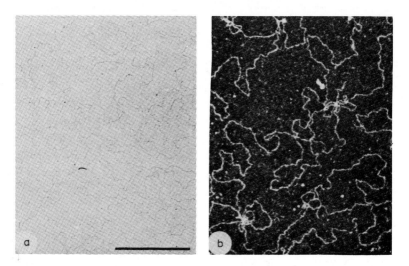

Fig. 5.15.   Electron micrograph of unstained λ-phage DNA: (a) bright-field, (b) dark-field (tilt). Image bar = 1 μm. (From Kleinschmidt 1971.)

and (b) dark-field (tilt) images of λ-phage DNA (Kleinschmidt 1971). For measurement of the length of the DNA strands, the dark-field image is preferable.

One of the reasons for developing high-voltage electron microscopes ($E_0 = 1$–3 MeV) was for the examination of whole unstained micro-organisms, such as bacteria and their flagella. However, because of the relatively weak interaction of high energy electrons with unstained biological material, small structures (10–50 nm) are imaged with poor contrast in bright-field. Figure 5.16a is a bright-field image of a bacterium (*Proteus* species) taken in a high-voltage microscope (Perrier 1973); although the bacteria are observed,

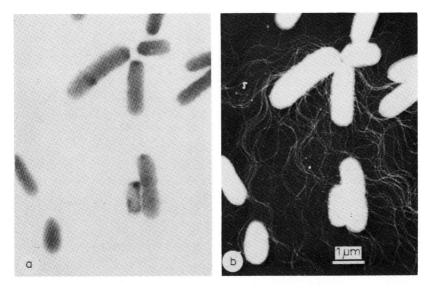

Fig. 5.16. Bacterium (*Proteus* species) in (a) bright-field and (b) dark-field. The edges are invisible in bright-field and therefore the bacteria cannot be seen to touch each other. Image bar = 1 μm, E_0 = 1 MeV. (From Perrier 1973.)

the flagella are not visible. The dark-field image (Fig. 5.16b) clearly shows these flagella, although the outlines of the bacteria are not well defined.

5.4.5 Atomic and molecular resolution

When the application of dark-field microscopy to high-resolution imaging is considered the results are rather controversial. This is because of two factors; namely, the problems of image interpretation and the increased radiation damage as compared with bright-field microscopy.

In a study of the ultrastructure of the iron core of the protein ferritin (Massover and Cowley 1973), the bright-field image (Fig. 5.17a) showed some internal structure of the iron core of the molecule, but most of the fine structure is similar to the background granular structure of the carbon support film. The dark-field (tilt) image (Fig. 5.17b) showed evidence of a crystallite structure in the iron core. Depending on the orientation of these crystallites with respect to the incident electron beam, certain cores showed an ordered structure, which corresponded to lattice planes separated by 0.95 nm. Massover and Cowley (1973) were aware of the possible misinterpretation of such images, such as the inability to distinguish between iron

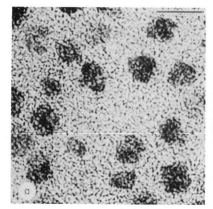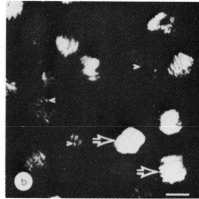

Fig. 5.17.   (a) Bright-field and (b) dark-field (tilt) images of unfractioned ferritin on an ultra-thin carbon support film. In the dark-field image most cores appear to have no ordered structure; scattering points less than 0.5 nm in diameter are present in some of these cores (arrowheads). Two cores consist of a single large crystal (arrows). Image bar = 10 nm, E_0 = 100 keV. (From Massover and Cowley 1973.)

deficient (low density) and iron rich (high density) regions of the ferritin molecule, and so X-ray diffraction results were used to confirm their interpretation of the dark-field images. Here radiation damage is relatively unimportant, because it is the relatively stable iron core that is being examined, and the disintegration of protein surrounding the core is probably not important.

The most spectacular results of the application of dark-field microscopy to imaging atoms and small molecules have been obtained by Ottensmeyer and co-workers. Detecting single atoms in organo-metallic compounds (Whiting and Ottensmeyer 1972) or in labelled nucleic acids (Henkelman and Ottensmeyer 1971) is an exercise in selecting from the micrograph dense regions corresponding to atoms amidst a noisy background; the selection is influenced by the known stereochemistry of the structures. If searching for heavy atoms labelling a specific base in a nucleic acid, the selection will be influenced by the fact that the atoms should form part of a continuous strand.

The 'microtephroscopy' (microscopy of the ashes) of small biological molecules assumes that, although the radiation dose is large, the damage to the molecules does not destroy their three-dimensional conformation (Ottensmeyer et al. 1975a, 1975b, 1977). For the protein myokinase, molecules selected from a noisy image were shown to be consistent with the known X-ray structure of the molecule (Fig. 5.18). One criticism of the technique is

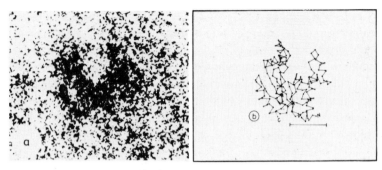

Fig. 5.18. Comparison of myokinase structure observed in dark-field microscopy with the structure determined by X-ray crystallography. Image bar = 2 nm, E_0 = 80 keV. (From Ottensmeyer et al. 1975b.)

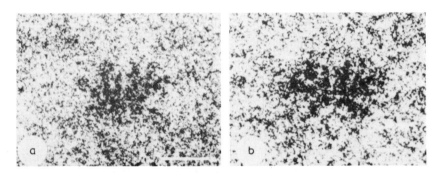

Fig. 5.19. Two dark-field micrographs of protamine, virtually identical in every detail, showing structure with measurable 0.5 nm image resolution; (a) from fraction Y–I, taken at 100,000 ×, (b) from fraction Y–I taken at 50,000 ×. Since the final magnification is the same, the difference in electron-optical magnification accounts for the difference in granularity of the images. Image bar = 2 nm, E_0 = 80 keV. (From Ottensmeyer et al. 1975b.)

that it is not possible to distinguish easily between genuine molecules, contaminants, or noise in the image. Figure 5.19 illustrates one test that should always be made, that is, recording two successive micrographs at different magnifications and preparing photographic prints so that the overall magnifications are the same; noise alters in this procedure, whereas genuine structures are similar. From images, such as those in Fig. 5.19, Ottensmeyer et al. (1975a) constructed a three-dimensional model of protamine, assisted by the known amino acid sequence of the protein together with other biochemical information. The application of dark-field microscopy to the determination of molecular structure is only viable for small (1–2 nm) unstained molecules, which may retain their structural integrity even after severe radiation damage. The technique is less likely to be applicable to large three-dimensional structures, such as virus subunits and ribosomes, which tend to collapse on the specimen grid in a configuration that is unlikely to resemble the original three-dimensional structure. Adequate control experiments should always be performed, so as to avoid the possibility of selecting visually attractive structures which are false. For example, separate specimen grids should be made with the unknown structure and with, for example, fragments of nucleic acids. Shapes should be selected from micrographs or prints by unbiased observers for both types of specimen, and a statistical estimate made of the success in selecting molecules from the correct specimen grid.

5.5   Coherent versus incoherent electron illumination

The electron optics of the illumination system in the electron microscope are normally designed to produce a well collimated, and hence, a coherent electron beam. The manufacturers assume that the operator will be using phase contrast to image thin specimens, and the electron source coherence is therefore made as high as possible to obtain the maximum amount of constructive and destructive interference in the image. However, it is this interference that causes most of the interpretation problems (such as phase contrast reversals with varying defocus) in bright-field microscopy. Enhancing the contrast of biological specimens by phase contrast also enhances the background structure, so that it is impossible to distinguish between fine structure arising from the specimen and its carbon support film (e.g. Fig. 5.17a). Of course, in bright-field microscopy it is possible to use image analysis and image processing to eliminate false detail in the image. However, if incoherent imaging conditions are used, the image is completely free from interference effects, and essentially the image represents true density variations in the specimen.

Incoherent imaging conditions are obtained when the illumination angle β is comparable to the objective aperture semi-angle α (Fig. 5.1; § 5.1). The degree of coherence between electrons scattered from two points r_1 and r_2 in the specimen is defined by the coherence function (Barnett 1974):

$$\Gamma(r_1, r_2) = \frac{2J_1\left(\dfrac{2\pi}{\lambda}\beta|r_1 - r_2|\right)}{\dfrac{2\pi}{\lambda}\beta|r_1 - r_2|} \tag{5.11}$$

where J_1 is the Bessel function of first order. If the value of $\Gamma(r_1, r_2)$ is near to unity for two points separated by $|r_1 - r_2|$, then the two points are coherently illuminated and electrons scattered from r_1 and r_2 will interfere. However, points separated by a distance $d_{coh} = |r_1 - r_2|$, such that the value of J_1 is zero, will be incoherently illuminated, and the electrons scattered from r_1 and r_2 will be imaged independently. This distance d_{coh} can be used to estimate an upper limit for the size of the coherence patch; points in the specimen separated by distances less than d_{coh} will be coherently illuminated, whereas points separated by distances greater than d_{coh} will be incoherently illuminated. d_{coh} then corresponds to the first zero of $J_1(X)$ at $X = 3.83$; thus:

$$\frac{2\pi}{\lambda} \beta d_{coh} = 3.83$$

or

$$d_{coh} = 0.61\lambda/\beta \tag{5.12}$$

Note that this is an upper limit, and a more stringent coherence requirement corresponds to the restriction of the value of $\Gamma(r_1, r_2)$ to between 1 and 0.88 (Barnett 1974), or $d_{coh} = 0.16\lambda/\beta$. The normal value for β is about 1 mrad, so that $d_{coh} = 2.3$ nm using Eq. (5.12). In order to obtain incoherent imaging conditions, β should be increased to 10 mrad by the combined use of a large second condenser aperture and the pre-field of the objective lens as a third condenser lens (Nagata et al. 1975, 1976). Nagata et al. (1976) observed that the phase contrast structure in a carbon film disappeared for a value of β of about 4 mrad. Whether it is desirable to use coherent or incoherent illumination can only be decided by practical examples, such as that shown in Fig. 5.20 (Nagata et al. 1975). The first pair of images (a) and (b) of platinum–palladium particles on a carbon film correspond to high resolution microscopy. It is very difficult to choose particles unambiguously in the coherent bright-field image (Fig. 5.20a), whereas, although the incoherent bright-field image (Fig. 5.20b) appears blurred and noisy, metal particles can be clearly seen. The second example (Fig. 5.20c and d) is a stained section imaged under (c) coherent and (d) incoherent imaging conditions; fine structure in the cell wall (Fig. 5.20c) is clearly a result of phase contrast and it cannot be believed. The detail visible in the incoherent bright-field image (Fig. 5.20d) is poor, but it is a more reliable representation of the cell wall structure.

Provided image processing facilities are available it is recommended that

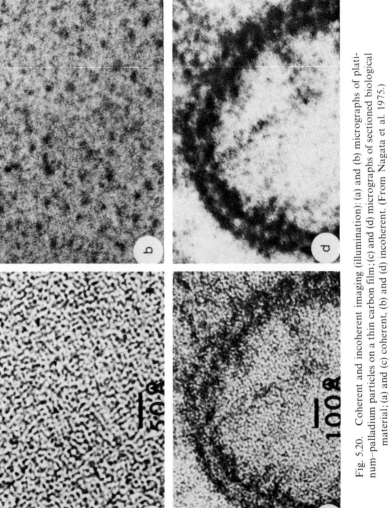

Fig. 5.20.   Coherent and incoherent imaging (illumination): (a) and (b) micrographs of platinum–palladium particles on a thin carbon film; (c) and (d) micrographs of sectioned biological
material; (a) and (c) coherent, (b) and (d) incoherent. (From Nagata et al. 1975.)

the advantage of phase contrast, namely, the linear relationship between j and η for thin specimens, is used. A high degree of source coherence is obtained by using a small (100 μm) second condenser aperture and working well away from cross-over. Such images can be corrected for phase contrast reversals and, for an ordered specimen, most of the background structure can be eliminated by selective filtering of the image transform (§ 4.2). The extent of the image transform and the attenuation of the higher spatial frequencies in the transform give some indication of the source coherence (§ 3.3.2; Fig. 3.9 and Fig. 3.10); by examining optical diffraction patterns for different illumination conditions, it is possible to find experimentally the conditions for maximum coherence (Beorchia and Bonhomme 1974).

In dark-field microscopy incoherent imaging conditions using hollow-cone illumination are recommended, so as to avoid the image interpretation problems associated with both tilt and axial dark-field images, which arise partly from the use of coherent illumination. The normal method of forming images in the STEM with an annular detector corresponds to dark-field incoherent imaging.

5.6   Optical shadowing or topographical imaging

An image formed by using the objective aperture to intercept one half of the diffraction pattern, but allowing the unscattered beam to contribute to the image, has the appearance of a heavy metal-shadowed specimen (Haydon and Lemons 1972) in which the 'angle of shadowing' depends on the orientation of the objective aperture with respect to the diffraction pattern. Such a *half-plane image* can be produced either by tilting the illumination so that the unscattered beam passes just inside the aperture, or a special D-shaped aperture with an indentation at its centre (to allow the unscattered beam to pass unattenuated) can be used. It is important not to intercept a significant part of the intense zero order (unscattered) beam otherwise severe electrical charging may occur at the objective aperture. Half-plane images will always have an astigmatic appearance, because of the use of only one side of the diffraction pattern, and it is very difficult to correct astigmatism with the half-plane aperture in place. There are two reasons for using this technique; firstly, for a thin specimen (weak phase object), the image transform does not have any frequency gaps such as occur in a normal bright-field image (§ 3.2.2). Secondly, half-plane images emphasize regions of the specimen where η may be small, but is varying rapidly. However, the interpretation of such images must be made with caution, because the enhancement of features in the specimen will depend on the orientation of the objective aperture with respect to the diffraction pattern (Krakow et al. 1976).

5.6.1   Theory of optical shadowing for thin specimens

In the weak phase approximation the object wave function $\psi_0(r) = 1 - i\eta(r)$, where $\eta(r)$ is the phase shift in the transmitted electron wave resulting from its interaction with the specimen (§ 3.2). The object transform or diffracted wave is $\delta - iA(v)$, and now the objective aperture is used to selectively filter one half of the two-dimensional transform $A(v) = A(v_x, v_y)$ (Scales 1974), with the objective aperture arranged, for example, so that the negative half-plane of A, corresponding to $v_x < 0$, is intercepted. The removal of one half of the diffracted wave allows an image to be formed that contains contributions from both the transforms $\sin \gamma$ and $\cos \gamma$; the equation for the image intensity $j(r)$ for a weak phase object is then:

$$j(r) = 1 - q(r) * \eta(r) + q'(r) * \frac{1}{\pi} \int \frac{\eta(x', y)}{x' - x} \, \mathrm{d}x' \qquad (5.13)$$

where $q(r)$ and $q'(r)$ are, respectively, the Fourier transforms of $\sin \gamma$ and $\cos \gamma$. The final term in Eq. (5.13) can be simplified, because the integration $\int (\eta(x', y)/(x' - x)) \, \mathrm{d}x'$ is *approximately* the differential of η in the x-direction, $\partial \eta / \partial x$, where the x-direction corresponds to the direction of 'shadowing'.

It will be evident that the image transform derived from Eq. (5.13) will contain a term in $\sin \gamma$ and a term in $\cos \gamma$. Thus whenever $\sin \gamma$ is small or zero, $\cos \gamma$ will have its maximum value, so that the image transform has no contrast transfer gaps. This absence of frequency gaps in the image transform of a half-plane image is shown in Fig. 5.21, which illustrates the transition from normal bright-field (left-hand side) to a half-plane image (right-hand side). The characteristic $\sin \gamma$ of the bright-field image becomes a uniform disc, with a small residual ring pattern along the aperture edge; the residual normal bright-field effect cannot be completely eliminated because the unscattered beam cannot be moved too near to the edge of the aperture without causing electrical charging problems. Thus for image processing the half-plane image has an ideal transfer function. Astigmatism and electrical charging of the objective aperture appear as distortions of this uniform disc, but it is not possible to determine the value of the defocus Δf from such an image transform, unless the residual normal bright-field ring pattern is used.

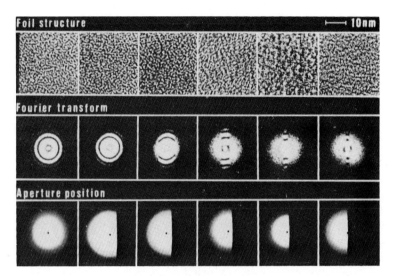

Fig. 5.21. The gradual transition from normal bright-field to a half-plane image for a thin carbon film, showing the image, its optical transform and the position of the objective aperture relative to the main beam. Image bar = 10 nm, $E_0 = 100$ keV, $\Delta f = 260$ nm. (P. Sieber, unpublished.)

In most images the value of η will be smaller than the differential of η, $\partial \eta / \partial x$, so that a half-plane image will emphasize structural features in the direction of shadowing. Figure 5.22 is a half-plane image of a negatively-stained T4 bacteriophage; the periodic tail structure is more evident than in normal bright-field images (see for example Fig. 3.28 in § 3.6.2).

According to the analysis above, half-plane images essentially represent the differential variations in η. Thus the technique is quite different from heavy metal shadowing, which emphasizes the surface topography of the specimen. The density variations in the half-plane images are a result of variations in η *through* the thickness of the specimen, and these density variations are not just a result of variations in surface topography. Consequently, although half-plane images appear to be similar to images of shadowed specimens, their interpretation may be quite different. High surface features of the specimen may correspond to regions of high values of the gradient $\partial \eta / \partial x$, but this may not always be true. The technique is virtually identical to the Schlieren method, which is used, for example, to enhance *optical density* gradients in liquids (Van Holde 1971).

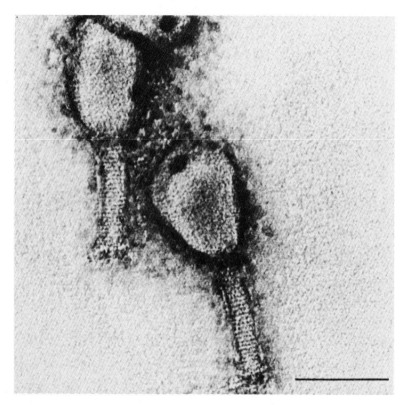

Fig. 5.22.   Half-plane (optically shadowed) image of a negatively-stained bacteriophage. Image bar $= 50$ nm, $E_0 = 100$ keV. (W.H.J. Andersen, unpublished.)

5.7   In situ *double exposures for enhancing phase contrast*

One problem in imaging biological specimens in bright-field is obtaining a uniform phase contrast transfer function over a large spatial frequency range (§ 3.2.2). An underfocus value of about 100 nm will give good phase contrast for spacings smaller than 1 nm, but poor phase contrast for larger spacings (Fig. 3.3; § 3.2.2). Using large underfocus values of about 500 nm will enhance large spacings around 2 nm, but give contrast reversals for spacings smaller than 1 nm (Fig. 3.4, § 3.2.2). In principle, it is possible to select from the image transforms of a focus series only parts of the transforms for which $\sin \gamma \simeq 1$, and combine them to give a perfect phase contrast image (§ 4.4). However, if image processing facilities are not available, the method given by Johansen (1975) for obtaining expanded phase contrast images *in situ* is a viable alternative. Since the relationship between image intensity *j* and the phase shift η is linear (Eq. (3.23); § 3.2), bright-field images can be added to give a composite bright-field image, whose transform is simply the addition of the two composite image transforms. Thus as shown in Fig. 5.23,

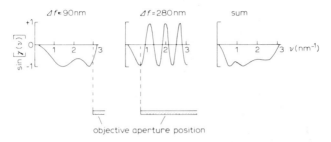

Fig. 5.23. The principle of double exposure phase contrast microscopy, with complementary objective apertures (shaded). The phase contrast transfer functions for $\Delta f_1 = 90$ nm (objective aperture diameter 35 μm) and for $\Delta f_2 = 280$ nm (objective aperture diameter 10 μm) are shown. The sum of the two contrast transfer functions gives the contrast transfer function of the superimposed images (Δf_1 and Δf_2). The contrast transfer function of the superimposed images gives an expanded phase contrast transfer region. (From Johansen 1975.)

the addition of two images with transfer functions appropriate to $\Delta f_1 = 90$ nm and $\Delta f_2 = 280$ nm gives an image with a transfer function that is the sum of these two transfer functions. Note that the transfer function for $\Delta f_2 = 280$ nm is cut off by using a smaller objective aperture before the transfer function decreases and changes sign. The use of the same size objective aperture in both cases would not be successful because the oscillations of $\sin \gamma_2$ would cancel with certain parts of $\sin \gamma_1$, and the resulting transfer function would be unsatisfactory. The contrast transfer function representing the sum of $\sin \gamma_1$ and $\sin \gamma_2$ (cut off before its first zero) gives a more extensive uniform phase contrast region than $\sin \gamma_1$ alone.

The appropriate objective aperture sizes are calculated from the spatial frequency range required, using $v_{max} = \alpha/\lambda$ with $\alpha = d_{obj}/2f_{obj}$. Thus for the image Δf_1 with $v_{max} = 3$ nm^{-1}, the objective aperture diameter d_{obj} should be approximately 36 μm ($f_{obj} = 1.6$ mm, $\lambda = 3.7$ pm), whilst for image Δf_2 with $v_{max} = 0.8$ nm^{-1}, d_{obj} will be 10 μm. So one image of the specimen is recorded with the smaller objective aperture (10 μm) in place and the objective lens defocused by approximately $\Delta f_2 = 280$ nm (with reference to the in-focus position of about 30 nm underfocus; § 3.2.2). The objective aperture mechanism is usually accurate enough for the larger objective aperture (35–40 μm) to be inserted in its correct position without the necessity of switching to diffraction and realigning the aperture. The second image is then recorded at an underfocus value of about 90 nm (Δf_1) on the same photographic film or plate as the first image. Evidently the two most serious experimental problems are the increased radiation damage to the specimen from recording two images and the possibility of specimen drift over the length of time required to correctly defocus and record two images. However, as illustrated in Fig. 5.24, the success in taking two such images can be verified by examin-

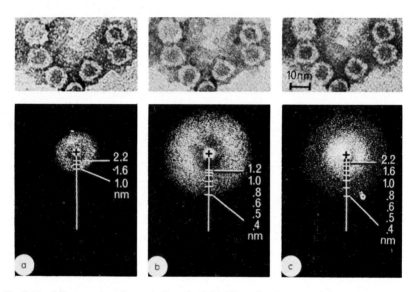

Fig. 5.24.   Micrographs of negatively-stained ferritin molecules, recorded at (a) 280 nm and (b) 90 nm underfocus. (c) is an *in situ* superimposed double exposure of the two focal settings in (a) and (b). Note the expanded range of spatial information in the optical transform of (c), 0.4–2.2 nm. $E_0 = 100$ keV. (From Johansen 1975.)

ing the optical diffraction pattern of the composite image. Johansen (1975) compared the composite image of negatively-stained ferritin and the corresponding optical transform (Fig. 5.24c) with the optical transforms of images corresponding to $\Delta f_2 = 280$ nm (Fig. 5.24a) and $\Delta f_1 = 90$ nm (Fig. 5.24b); note that the scale marked on the optical transforms is in terms of the spacing r (nm) and not spatial frequency $v = 1/r$ (nm^{-1}). Although the composite image (Fig. 5.24c) does have a more extensive region of uniform contrast transfer than either of the original images, it does not show any significant additional detail. The same is true of the images of the negatively-stained T2 bacteriophage (Fig. 5.25), although the tail fibres are more

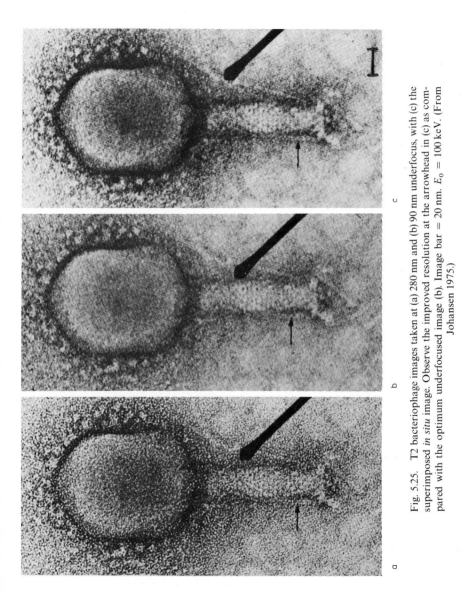

Fig. 5.25. T2 bacteriophage images taken at (a) 280 nm and (b) 90 nm underfocus, with (c) the superimposed *in situ* image. Observe the improved resolution at the arrowhead in (c) as compared with the optimum underfocused image (b). Image bar = 20 nm. $E_0 = 100$ keV. (From Johansen 1975.)

evident in the composite image (Fig. 5.25c). This technique may not be used to its full potential on negatively-stained specimens, because most of the significant image details correspond to spacings larger than 2 nm.

A similar result to the *in situ* superposition of two images can also be made photographically using images taken at the appropriate two defocus values; the problem is then to align the images accurately by eye. The photographic superposition made on film can then be examined on an optical diffracto-meter; the result of misaligned images will be evident as a series of fringes in the optical transform.

References

Agar, A.W., R.H. Alderson and D. Chescoe (1974), Principles and practice of electron micro-scope operation, in: Practical methods in electron microscopy, Vol. 2, A.M. Glauert, ed. (North-Holland, Amsterdam).

Barnett, M.E. (1974), Image formation in optical and electron transmission microscopy, J. Microsc. (Oxford) *102*, 1–28.

Beorchia, A. and P. Bonhomme (1974), Experimental studies of some dampings of electron microscope phase contrast transfer functions, Optik *39*, 437–442.

Burge, R.E. (1973), Mechanisms of contrast and image formation of biological specimens in the transmission electron microscope, J. Microsc. (Oxford) *98*, 251–285.

Cowley, J.M. (1975a), Coherent and incoherent imaging in the scanning transmission electron microscope, J. Phys. D: Appl. Phys. *8*, L77–79.

Cowley, J.M. (1975b), A comparison of scanning and fixed beam high-voltage electron micro-scopy, in: Physical aspects of electron microscopy, B.M. Siegel and D.R. Beaman, eds. (John Wiley and Sons, New York), pp. 17–28.

Cowley, J.M. (1975c), Contrast in high-resolution bright-field and dark-field images of thin specimens, in: Physical aspects of electron microscopy, B.M. Siegel and D.R. Beaman, eds. (John Wiley and Sons, New York), pp. 3–15.

Cowley, J.M., W.H. Massover and B.K. Jap (1974), The focusing of high resolution dark field electron microscope images, Optik *40*, 42–54.

Crewe, A.V. (1971), High resolution scanning microscopy of biological specimens, Phil. Trans. Roy. Soc. Lond. B *261*, 61–70.

Crewe, A.V. and J. Wall (1970), A scanning microscope with 5 Å resolution, J. Mol. Biol. *48*, 375–393.

Crewe, A.V., J.P. Langmore and M.S. Isaacson (1975), Resolution and contrast in the scanning transmission electron microscope, in: Physical aspects of electron microscopy, B.M. Siegel and D.R. Beaman, eds. (John Wiley and Sons, New York), pp. 47–62.

Cullis, A.G. and D.M. Maher (1974), High-resolution topographical imaging by direct trans-mission electron microscopy, Phil. Mag. *30*, 447–451.

Cullis, A.G. and D.M. Maher (1975), Topographical contrast in the transmission electron microscope, Ultramicroscopy *1*, 97–112.

Dubochet, J. (1973), High resolution dark-field electron microscopy, J. Microsc. (Oxford) *98*, 334–344,

Dupouy, G. (1967), Contrast improvement in electron microscopic images of amorphous objects, J. Electron Microsc. *16*, 5–16.

Dupouy, G., F. Perrier, L. Enjalbert, L. Lapchine and P. Verdier (1969), Accroissement du contraste des images d'objets amorphes en microscopie électronique, C.R. Acad. Sci. Paris B *268*, 1341–1345.

Egerton, R.F. (1976), Inelastic scattering and energy filtering in the transmission electron microscope, Phil. Mag. *34*, 49–65.

Egerton, R.F., J.G. Philip, P.S. Turner and M.J. Whelan (1975), Modification of a transmission electron microscope to give energy-filtered images and diffraction patterns, and electron energy loss spectra, J. Phys. E: Sci. Instrum. *8*, 1033–1037.

Engel, A., J.W. Wiggins and D.C. Woodruff (1974), A comparison of calculated images generated by six modes of transmission electron microscopy, J. Appl. Phys. *45*, 2739–2747.

Engel, A., J. Dubochet and E. Kellenberger (1976), Some progress in the use of a scanning transmission electron microscope for the observation of biomacromolecules, J. Ultrastruct. Res. *57*, 322–330.

Gentsch, P., H. Gilde and L. Reimer (1974), Measurement of the top bottom effect in scanning transmission electron microscopy of thick amorphous specimens, J. Microsc. (Oxford) *100*, 81–92.

Goodman, J.W. (1968), An introduction to Fourier optics (McGraw-Hill, New York).

Hanszen, K.-J. (1969), Problems of image interpretation in electron microscopy with linear and nonlinear transfer, Z. Angew. Phys. *27*, 125–131.

Hanszen, K.-J. (1973), Contrast transfer and image processing, in: Image processing and computer-aided design in electron optics, P.W. Hawkes, ed. (Academic Press, London), pp. 16–53.

Haydon, G.B. and R.A. Lemons (1972), Optical shadowing in the electron microscope, J. Microsc. (Oxford) *95*, 483–491.

Henkelman, R.M. and F.P. Ottensmeyer (1971), Visualization of single heavy atoms by dark field electron microscopy, Proc. Natl. Acad. Sci. USA *68*, 3000–3004.

Henkelman, R.M. and F.P. Ottensmeyer (1974), An energy filter for biological electron microscopy, J. Microsc. (Oxford) *102*, 79–94.

Isaacson, M. and D. Johnson (1975), The microanalysis of light elements using transmitted energy loss electrons, Ultramicroscopy *1*, 33–52.

Isaacson, M., J.P. Langmore and H. Rose (1974), Determination of the non-localization of the inelastic scattering of electrons by electron microscopy, Optik *41*, 92–96.

Johansen, B.V. (1975), Bright field electron microscopy of biological specimens – III. Expanded phase contrast information using *in situ* double exposures with complementary objective apertures, Micron *6*, 153–163.

Johnson, H.M. and D.F. Parsons (1969), Enhanced contrast in electron microscopy of unstained biological material – I. Strioscopy (dark-field microscopy), J. Microsc. (Oxford) *90*, 199–220.

Kleinschmidt, A.K. (1971), Electron microscopic studies of macromolecules without apposi-tional contrast, Phil. Trans. Roy. Soc. Lond. B *261*, 143–149.

Krakow, W., D.G. Ast, W. Goldfarb and B.M. Siegel (1976), Origin of the fringe structure observed in high resolution bright-field electron micrographs of amorphous materials, Phil. Mag. *33*, 985–1014.

Lovell, D.J. and S.K. Chapman (1975), The effect of variations in objective focal length on electron microscope performance, J. Microsc. (Oxford) *105*, 277–282.

Massover, W.H. and J.M. Cowley (1973), The ultrastructure of ferritin macromolecules. The lattice structure of the core crystallites, Proc. Natl. Acad. Sci. USA *70*, 3847–3851.

Misell, D.L. (1976), Contrast enhancement by changing the objective lens focal length, J. Microsc. (Oxford) *108*, 13–20.

Misell, D.L. (1977), Conventional and scanning transmission electron microscopy: image contrast and radiation damage, J. Phys. D: Appl. Phys. *10*, 1085–1107.

Misell, D.L. and I.D.J. Burdett (1977), Determination of the mass thickness of biological sections from electron micrographs, J. Microsc. (Oxford) *109*, 171–182.

Misell, D.L., G.W. Stroke and M. Halioua (1974), Coherent and incoherent imaging in the scanning transmission electron microscope, J. Phys. D: Appl. Phys. *7*, L113–117.

Nagata, F., T. Matsuda and T. Komoda (1975), High resolution electron microscopy by an incoherent illumination method, Japan. J. Appl. Phys. *14*, 1815–1816.

Nagata, F., T. Matsuda, T. Komoda and K. Hama (1976), High resolution observation of biological specimens by an incoherent illumination method, J. Electron Microsc. *25*, 237–243.

Ottensmeyer, F.P. and M. Pear (1975), Contrast in unstained sections: A comparison of bright and dark field electron microscopy, J. Ultrastruct. Res. *51*, 253–260.

Ottensmeyer, F.P., R.F. Whiting and A.P. Korne (1975a), Three-dimensional structure of herring sperm protamine Y–I with the aid of dark field electron microscopy, Proc. Natl. Acad. Sci. USA *72*, 4953–4955.

Ottensmeyer, F.P., R.F. Whiting, E.E. Schmidt and R.S. Clemens (1975b), Electron micro-tephroscopy of proteins: A close look at the ashes of myokinase and protamine, J. Ultrastruct. Res. *52*, 193–201.

Ottensmeyer, F.P., J.W. Andrew, D.P. Bazett-Jones, A.S.K. Chan and J. Hewitt (1977), Signal-to-noise enhancement in dark field electron micrographs of vasopressin: filtering of arrays of images in reciprocal space, J. Microsc. (Oxford) *109*, 259–268.

Perrier, F. (1973), Aspects of dark-field electron microscopy, J. Microsc. (Oxford) *98*, 352–358.

Reimer, L. and P. Gentsch (1975), Superposition of chromatic error and beam broadening in transmission electron microscopy of thick carbon and organic specimens, Ultramicroscopy *1*, 1–5.

Scales, D.J. (1974). A Fourier approach to optical shadowing with a transmission electron microscope, J. Microsc. (Oxford) *102*, 49–58.

Stroke, G.W. and M. Halioua (1972), Attainment of diffraction-limited imaging in high-resolution electron microscopy by *a posteriori* holographic image sharpening, I, Optik *35*, 50–65.

Thomson, M.G.R. (1973), Resolution and contrast in the conventional and the scanning high resolution transmission electron microscopes, Optik *39*, 15–38.

Thon, F. and D. Willasch (1972), Imaging of heavy atoms in dark-field electron microscopy using hollow cone illumination, Optik *36*, 55–58.

Tschopp, J. and P.R. Smith (1977), Extra-long bacteriophage T4 tails produced under *in vitro* conditions, J. Mol. Biol. *114*, 281–286.

Van Holde, K.E. (1971), Physical biochemistry (Prentice-Hall, Inc., Engelwood Cliffs, New Jersey), Chapter 5.

Whiting, R.F. and F.P. Ottensmeyer (1972), Heavy atoms in model compounds and nucleic acids imaged by dark-field transmission electron microscopy, J. Mol. Biol. *67*, 173–181.

Chapter 6

Computer methods
of contrast enhancement

This chapter describes methods of contrast enhancement that do not use Fourier techniques, but which are applied directly to a digitised micrograph.

It is standard practice when printing micrographs to choose the photographic emulsion of the paper or film and their developing conditions so as to produce a high contrast result; the original density or grey levels in the micrograph of a stained specimen are altered in the print so that high density levels (stain; low density levels on the original micrograph) are enhanced with respect to the lower density levels (background or unstained material; high density levels on the original micrograph). This photographic enhancement (e.g. Farnell and Flint 1970, 1973) should be interpreted with caution because it represents a non-linear process, in which the original linear relation between the image intensity j and some property of the specimen may be altered. A digitised micrograph can also be processed in this way: the original density-exposure characteristics of the micrograph (Agar et al. 1974) are altered by multiplying each picture element, or *pixel*, by some factor that effectively enhances the low density values and reduces high density values in the original micrograph. Note that a final print of an image will have reversed contrast with respect to the micrograph, so that dense regions of the specimen (e.g. stain) that appear light in the micrograph, are dark in the print. One simple way to enhance contrast in the image is to subtract the background (§ 6.1). This can be readily achieved on a digitised image, but cannot be done photographically because optical densities can only be added.

The enhancement of boundary or edge structures is possible because of the marked variations in the optical density that occur near a boundary (e.g. a cell wall) or edge (e.g. the demarcation between a biological structure and its surrounding negative stain). Although the image contrast may be low, the optical density variations near the edge may be quite large, and, by selecting only those elements of the image that correspond to large density variations, it is possible to reduce the effect from the unstructured background, where the density variations are small (§ 6.3). The distribution of density levels in the image is often characteristic of particular types of structure; for example, low density values in the image of a stained section will be characteristic of positively-stained structures whilst high density values will be appropriate to the embedding medium and unstained biological material. Thus it may be possible to *threshold* the digitised image so that only picture elements of low density (stain) are retained, and picture elements of high density are omitted. In this way stained regions of the specimen are enhanced (§ 6.3.1). It is often difficult to decide where the demarcation between 'structure' and background is in low contrast images of unstained sections; here the *Laplacian*, the second differential of the image, is used to distinguish between regions of the image where the optical density variations are large and regions where they are small (Weszka et al. 1974; § 6.3.2).

Image subtraction is a method for reducing the contribution of inelastic scattering to high resolution images (Misell and Burge 1975). The 'elastic' image is assumed to contain high resolution information and is superimposed on a low resolution 'inelastic' image, blurred by chromatic aberration (Cosslett 1956). Whereas the resolution of the elastic image is sensitive to defocus changes, the inelastic image is relatively unaffected by defocus changes of 50–100 nm (§ 6.2). Two images are taken, the first at the optimum defocus Δf_{opt} (§ 3.2) for the elastic image and the second at a defocus 50–100 nm away from the optimum. The first picture consists of a high resolution elastic image superimposed on an inelastic background, whilst the second is a low resolution elastic image on a virtually unchanged inelastic background. A subtraction of two such images, either by computer or photographically (Krakow et al. 1976), should produce an image with a significantly reduced inelastic background and an enhanced image of the high resolution elastic component. Because the inelastic component has very little effect on the contrast of bright-field images, the application of this type of image subtraction is exclusive to dark-field images, in which inelastic scattering may significantly reduce image contrast (see § 5.4.1; Table 5.2). Also image subtraction will not work with specimens of intrinsically low resolution, such as biological sections and negatively-stained specimens; for these specimens the elastic component of the image contains no useful structural information below 2 nm resolution and this is not significantly better than the resolution of the inelastic image (§ 2.5.2). An attempt to apply image subtraction to negatively-stained groups of nine hexons from adenovirus (see Fig. 2.4; § 2.4) produced a result that represented changes in the negative-stain pattern between the two successive exposures and eliminated most of the character-istic shape of the groups of nine hexons.

The technique of image subtraction should be applied to high resolution dark-field images, where the inelastic image is significantly inferior to the intrinsic resolution of the elastic image. It may be, for example, applied to the enhancement of images of heavy metal-labelled macromolecules and small unstained molecules (1–2 nm), which retain their structural integrity even after large radiation doses. The reduction in the contribution from the inelastic scattering to the image may assist the selection of genuine structures from a noisy electron micrograph.

6.1 Subtraction of background

The subtraction of background from an image is effected by subtracting the mean value of the optical density D_m of the digitised image from the density value D of each picture element. This is particularly useful for enhancing the contrast of bright-field micrographs which contain large background contri-butions from the unscattered component of the transmitted electron beam. In practice, the mean density value is calculated by adding the density values (usually arbitrarily scaled from 0 to 256) of all the picture elements ($N \times N$) and dividing this total by the total number of picture elements N^2. A new picture array is then generated with density values $D - D_m$; this range of density values is then divided in equal density increments according to the number of grey levels available on the television display system. In the example shown in Fig. 6.1 the television display has 9 grey levels: (a) is the digitised image of a stained section of *Bacillus subtilis*; (b) the result after background subtraction. The contrast has been enhanced by subtracting the background, but there is no more detail evident in the cell wall. Note also that the signal-to-noise ratio in the two images is unaltered by background subtraction; the noise is more evident in Fig. 6.1b, because it is no longer masked by the background.

In addition to background subtraction, the density values of the image can be altered so as to artificially enhance or reduce contrast; this is equivalent to altering photographically the slope or *gamma* of the density-exposure characteristic of the image (Agar et al. 1974). If contrast enhancement is required, it is usual to reverse the contrast of the digitised image before multiplying the density values by an exponential or similar factor. For

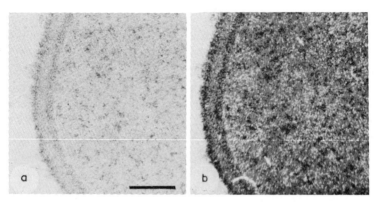

Fig. 6.1.   Background subtraction for a thin stained section of *Bacillus subtilis*. $\alpha = 1.5 \times 10^{-2}$
rad, $E_0 = 60\,\text{keV}$, image bar = 50 nm.

example, the original density values, 0–256, are reversed to 256–0, so that the original low density values correspond to high numbers. These modified density values can then be multiplied by the exponential factor $\exp(cD)$, which increases high D values relative to low D values; c is an arbitrary constant to be chosen (see below). In allocating grey levels on the display, the new density range is divided into the appropriate number of equidistantly-spaced grey levels. In this way low density values are compressed into a small number of grey levels, whilst high density values occupy most of the available grey levels, and so determine the appearance of the image to the eye. The modified image is usually photographed either by using a Polaroid camera or a 35 mm camera; the grey level scale on the television display immediately below the image is also photographed. The 35 mm film should be developed so that the grey level scale is linear and all grey levels are observable on the scale. Unfortunately there is no control over the developing conditions of Polaroid film, so it may be necessary to adjust the grey level scale on the display, or to take many pictures at varying exposures, to ensure linearity on the film. The constant c is chosen to give an appropriate enhancement of the high density values; for example, if the top density level of 256 ($D_{\max}$) is to be enhanced by a factor of 10 (f) with reference to the bottom density level of 0 ($D_{\min}$), c should be chosen to be equal to $\log_e f / (D_{\max} - D_{\min})$ $= \log_e 10/(256-0) = 0.009$. If the original image is divided into 9 equal grey levels, picture elements with density values in the range 0–142 will occupy 5 of the grey levels available, whereas in the modified image these same picture elements will be restricted to only 2 grey levels; the high density values 143–256 will occupy the remaining 7 grey levels and these picture elements will dominate the visual appearance of the image. The reduction in image contrast by taking the logarithm of the picture density values is unlikely to be of general interest. However, there is one example where such a procedure may be useful, namely a linear display of mass thickness variations ρt in a sectioned specimen (§ 5.1). The intensity distribution, I, in a scattering contrast image is related to ρt by Eq. (5.2): $I = I_0 \exp(-S_p \rho t)$ (§ 5.1). The logarithm of the image intensity $\log_e I = -S_p \rho t +$ an arbitrary constant, where this constant can be subtracted from the image. Then a display of $\log_e I$ will show linear variations in ρt.

6.2   Image subtraction

Image subtraction depends on the differential behaviour of the elastic and inelastic images with changes in focus (Misell and Burge 1975). Because of chromatic aberration, the inelastic image is out of focus relative to the elastic image by $\Delta f = C_c \Delta E / E_0$, which is approximately 500 nm for $E_0 = 100$ keV and an energy loss $\Delta E = 25$ eV ($C_c = 2$ mm) or, in terms of image resolution, each image point in the inelastic image will be blurred out to a diameter of about 2 nm as a result of chromatic aberration (Misell 1975). The resolution of the elastic image is mainly determined by spherical aberration and defocus. The resolution function for elastic scattering, which is the Fourier transform of the microscope transfer function (§ 3.2), measures how each point in the specimen will appear in the image; it depends critically on defocus and attains its best form near to optimum defocus, about 50 nm underfocus in dark-field microscopy (§ 5.4.2). The inelastic resolution function is virtually unaltered by focus changes of 50–100 nm.

6.2.1   Theory of image subtraction

Consider two electron micrographs of the same field, taken with identical electron exposures, with defocus values Δf_1 and Δf_2, respectively, where Δf_1 is optimum elastic defocus and Δf_2 differs from this optimum by 50–100 nm. The two image intensity distributions and the difference image intensity distribution are given by:

$$\left. \begin{aligned} \text{image 1}: \ & j_1(\mathbf{r}) = [j_E(\mathbf{r})]_1 + [j_I(\mathbf{r})]_1 \\ \text{image 2}: \ & j_2(\mathbf{r}) = [j_E(\mathbf{r})]_2 + [j_I(\mathbf{r})]_2 \\ \text{difference image} : \ & \Delta j(\mathbf{r}) = \Delta j_E(\mathbf{r}) + \Delta j_I(\mathbf{r}) \end{aligned} \right\} \qquad (6.1)$$

where $r = (x, y)$ defines the position coordinate in the image, and j_E and j_I refer respectively to the contributions to the total image intensity j (elastic plus inelastic) from the two types of scattering.

The success of subtraction in reducing the inelastic contribution depends on the form of the variations in the image profile with defocus. This was examined by calculating image point resolution functions for elastic and inelastic imaging in dark-field microscopy (using axial illumination and a dark-field central stop corresponding to a semi-angle $\alpha_{min} = 1$ mrad; § 5.4.2). The functions for (a) elastic and (b) inelastic imaging are shown in Fig. 6.2 for

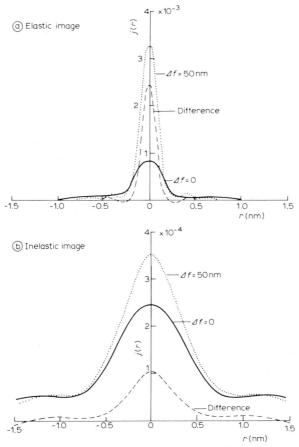

Fig. 6.2.   Illustration of image subtraction for coherent dark-field microscopy. The image profiles for both (a) elastic and (b) inelastic scattering are shown for $\Delta f = 0$ and $\Delta f = 50$ nm. The respective difference curves are shown by dashed lines. $E_0 = 100$ keV, $\alpha = 10$ mrad, $\alpha_{min} = 1$ mrad, $C_s = C_c = 2$ mm.

defocus values of 50 nm (optimum defocus) and 0 nm. The difference image profiles Δj_E and Δj_I are shown as dashed lines. There is a small reduction (20 %) in the contribution from elastic scattering but a much larger (70 %) reduction in the inelastic scattering. The difference image then comprises a slightly reduced elastic component on a significantly reduced inelastic background. Quantitatively, the gain in image contrast can be determined from the change in the relative proportions of elastic and inelastic scattering. In the original images, the values for the fraction of electrons scattered inelastically $I(\alpha)$ and elastically $E(\alpha)$ within an objective aperture of semi-angle 10 mrad for a carbon specimen of thickness 10 nm are respectively 0.100 and 0.016, and the image contrast before subtraction $C \simeq E(\alpha)/[E(\alpha) + I(\alpha)] = 14\%$. After image subtraction the elastic contribution is reduced by 24 % whereas the inelastic contribution is reduced by 72 %. Contrast is increased to 31 %, a substantial gain.

6.2.2 Practical problems in image subtraction

As in all methods requiring two images of the same area, radiation damage is a serious problem; the second image will be degraded. In dark-field microscopy there is the additional problem of taking two images at accurately determined defocus values (§ 5.4.2).

Image subtraction can be effected photographically or by computer. Since optical densities can only be added photographically, photographic image subtraction can only be implemented by superimposing the negative of one image on the positive of the second image (Krakow et al. 1976). Using computer methods the two images must be digitised and scaled so that the optical densities of each represent the same total number of electrons detected (§ 4.7.2). It is impossible to take two successive images which have exactly the same *total energy density*; that is, so that the total number of electrons detected by each plate/film is the same for each image. It is assumed for image subtraction that, although the spatial distribution of electrons in the images are different (as a result of their different resolutions), the total number of electrons transmitted by the specimen is the same. Thus the two image intensities $j_1(x, y)$ and $j_2(x, y)$ are normalised so that over a large region of each image:

$$\sum_x \sum_y j_1(x, y) = \sum_x \sum_y j_2(x, y) \qquad (6.2)$$

This normalisation is achieved by the separate addition of the image intensities for each image, and then multiplying one set of image intensities, say j_1, by the ratio of $\sum j_2/\sum j_1$. Photographically this type of scaling is not easy to implement without making several photographic copies of one image and evaluating the mean density levels using a microdensitometer.

The basic problem of the alignment of two images for subtraction is the translation and rotation of one image with respect to the second, so that the prominent features of each image coincide. This procedure may be done photographically by superposition of the negative of one image on the positive of the other; the choice of the position of best alignment is subjective, with the eye choosing the result that shows the best 'structure' with the least background. Additional problems occur in photographic subtraction as a result of the difficulty in matching the optical densities of the positive and negative of one image. It may be necessary to prepare several negatives of one image at different optical densities and to use several different film emulsions to match the optical density – exposure characteristics of the negative with the positive image (Farnell and Flint 1973). This is difficult to do because the characteristics of photographic emulsions are quite different for electrons (the original positive) and light (the negative copy). Computer alignment of two similar images has been considered in § 4.7.2; it is more objective than photographic alignment because the accuracy of alignment can be assessed from the width of the cross-correlation function, $c(x)$, as shown in Fig. 4.31 in § 4.7.2. Subtraction of the two aligned images is then effected by subtracting the density values of corresponding picture elements.

6.2.3   Applications of image subtraction

Figure 6.3 is an example of image subtraction by computer, applied to images of uranyl acetate-stained SV40 DNA, obtained using axial dark-field microscopy (Krakow et al. 1976). Figure 6.3a is the image taken at optimum defocus, whilst Fig. 6.3c is an image a further 200 nm out of focus; the difference image is shown in Fig. 6.3e. The sequence Fig. 6.3b, d, f are reversed-contrast versions of Fig. 6.3a, c, e. The width of the stained DNA strand in the difference image is a factor of two smaller than in either of the two images used for subtraction. This high resolution detail was masked by inelastic scattering in both of the original images. But since the images in Fig. 6.3 form part of a series of 22 images, it is unlikely that it is the DNA structure that has been resolved. It is most likely that subtraction has removed the predominantly inelastic components of the DNA damage products and the

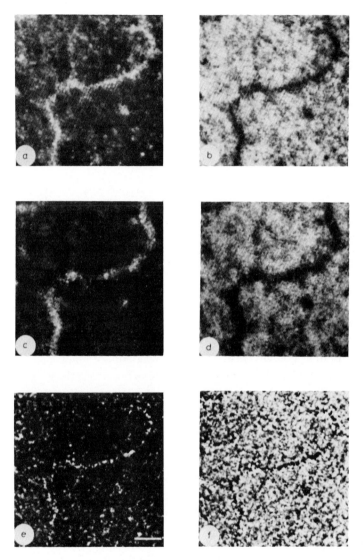

Fig. 6.3.   Image subtraction for uranyl acetate-stained SV40 DNA in dark-field (central beam stop): (a) best focus (b) best focus with reversed contrast, (c) 'inelastic' image, (d) 'inelastic' image with reversed contrast, (e) difference image (by computer), (f) difference image with reversed contrast. All images have 256 × 256 picture elements. Image bar = 20 nm, $E_0 = 100$ keV. (From Krakow et al. 1976.)

carbon support film to leave the uranium atoms, which form an essentially elastic image, in clear contrast. Image subtraction will therefore be most successful in resolving heavy atoms or groups of heavy atoms substituted in biological macromolecules. The technique may be useful in improving the poor image contrast obtained from small unstained molecules, provided that the ashes retain a three-dimensional structure with significant high resolution detail.

6.3   Enhancement of boundary structure

A boundary or an edge in the image is a zone separating two regions of low and high optical density; it is usually characterised by a large change in the optical density gradient. In electron microscopy, such a boundary may represent the demarcation between positively-stained and unstained biological material in a section, or delineate a biological molecule from its negative stain. Often this boundary is obscured by a background which lowers image contrast. In images of unstained sections it may be difficult to see any detail because of the poor contrast. In stained specimens it is often possible to assign certain optical density levels to stain and to form an image which has been *thresholded* so that only these density levels are retained (§ 6.3.1). If this type of thresholding is not possible, it may be possible to enhance a boundary by using the fact that there is usually a large optical density gradient across it; only these regions are retained (§ 6.3.2).

6.3.1   The density histogram and thresholding

Thresholding is made on the basis of the *histogram* of optical densities; that is, a graph of the number of picture elements with particular optical density values. Figure 6.4 shows the density histogram obtained from the digitised image of a stained section of *Bacillus subtilis* (Fig. 6.5a). The histogram resolves clearly into two peaks, the low density peak which represents regions of stain (dark in Fig. 6.5a) and a high density peak, the embedding medium and unstained material (light in Fig. 6.5a). Thus all picture elements with density values between 170 and 200 are retained in the image array; the background value of 170 is subtracted from this to give an image with density values between 0 and 30, which are assigned to 9 grey levels. This thresholded image is shown in Fig. 6.5c. The image corresponding to density values 200–220 (Fig. 6.5d) is virtually unstructured. This type of thresholding can only be used to enhance contrast when there are two

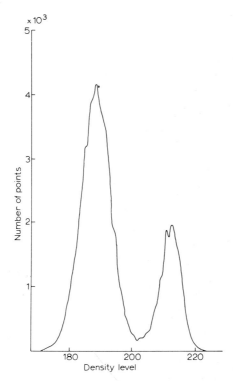

Fig. 6.4. Histogram of densities (arbitrary units) for a stained section of *Bacillus subtilis*. The high density peak corresponds to unstained material (background), and the low density peak to biological (stained) material. The division into two peaks makes thresholding possible.

distinct peaks in the density histogram of the image. Even in the example shown in Fig. 6.4 only about 25% of the image points occur in the second peak, so that thresholding does not produce a dramatic enhancement of the cell wall structure. The best results of thresholding based on the density histogram occur in examples where the 'background' peak constitutes more than 50% of the picture elements (Weszka et al. 1974).

For a low contrast image such as that obtained from an unstained section of *Bacillus subtilis* (osmium post-fixed) the density histogram shows only a single peak (Fig. 6.6), and from this alone it is not possible to separate the image into low and high density picture elements. The alternatives are to either differentiate the image to enhance regions with large density gradients or to use a different criterion for thresholding based on the *Laplacian* (second derivative) of the image (Weszka et al. 1974).

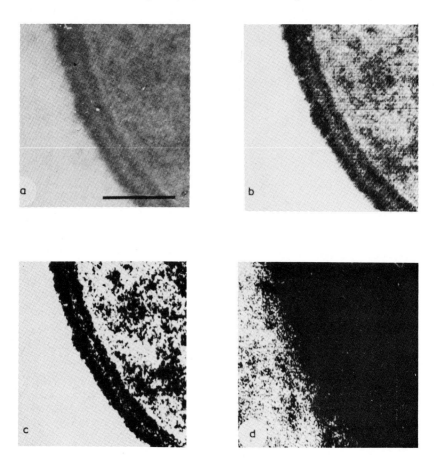

Fig. 6.5. (a) Stained section of *Bacillus subtilis* (glutaraldehyde/OsO$_4$-fixed and uranyl acetate/lead citrate-stained, (b) background subtracted, (c) histogram cut – low densities only (stain), (d) histogram cut – high densities only (support film, embedding medium). $E_0 = 60$ keV, image bar = 100 nm.

6.3.2   The use of the Laplacian for enhancing boundary structure

The optical differences between a picture element and its adjacent horizontal and vertical picture elements represents the rate of change, or first derivative, of the optical density in the x- and y-directions. The density gradient of a particular image point is then calculated as the square root of the sum of the squares of horizontal (x) and vertical (y) derivatives. Edges or boundaries in the image are therefore calculated without regard to their direction in the

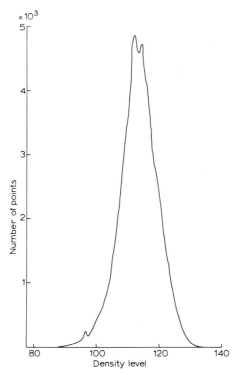

Fɪɢ. 6.6.   Histogram of densities (arbitrary units) for an unstained section of *Bacillus subtilis*. There is now only a single peak and the separation of background from biological structure is not possible.

image, and the resulting picture, generated using these gradients at each picture point, will show only those features in the original image that correspond to significant variations in density; background and relatively unstructured regions will be absent from the image.

The Laplacian method of edge enhancement is based on the second order derivative. It emphasizes maximum values or peaks in the image. Physiological research in human vision indicates that the eye sees objects in much the same way. The Laplacian at a particular image point is calculated by subtracting the density value of the picture point from the average density value of the same point and its eight neighbours. The Laplacian is generated by repeating this procedure for all picture elements. However, in both the gradient and Laplacian methods of edge enhancement, noise, which usually shows rapid changes in density, will also be enhanced, so that the final result of differentiation may be extremely noisy. This can be reduced either by averaging adjacent picture elements in the differentiated image or by calculating a derivative, such as the Laplacian, based on an average over, say, a 7×7 or an 11×11 square centred at the given image point, rather than the 3×3 square used above. This use of a 'coarse' Laplacian enables broad boundaries or edges to be detected, whereas the 'fine' Laplacian may generate high values originating from the background noise rather than from points near the edge or boundary (Weszka et al. 1974).

Although the Laplacian of the image may be noisy, it can be used as an objective test for selecting regions of the original image which contain edges or boundaries. Only the original picture elements with a Laplacian significantly different from zero are retained, so that regions of the original with small density variations are suppressed.

Edge enhancement and similar techniques of contrast enhancement have not yet been applied to micrographs, but they have been successfully applied to optical images. The application of the techniques described here and the results obtained are illustrated in an excellent booklet, 'Computer eye: Handbook of image processing', published by Spatial Data Systems, Inc. (see List of Suppliers).

References

Agar, A.W., R.H. Alderson and D. Chescoe (1974), Principles and practice of electron microscope operation, in: Practical methods in electron microscopy, Vol. 2, A.M. Glauert, ed. (North-Holland, Amsterdam).

Cosslett, V.E. (1956), Specimen thickness and image resolution in electron microscopy, Brit. J. Appl. Phys. 7, 10–12.

Farnell, G.C. and R.B. Flint (1970), A method for increasing the photographic contrast of electron micrographs, J. Microsc. (Oxford) 92, 145–151.

Farnell, G.C. and R.B. Flint (1973), The response of photographic materials to electrons, with particular reference to electron microscopy, J. Microsc. (Oxford) 97, 271–291.

Krakow, W., K.B. Wells and B.M. Siegel (1976), Image processing of dark-field electron micrographs, J. Phys. D: Appl. Phys. 9, 175–181.

Misell, D.L. (1975), The resolution and contrast in biological sections determined by inelastic and elastic scattering, in: Physical aspects of electron microscopy, B.M. Siegel and D.R. Beaman, eds. (John Wiley and Sons, New York), pp. 63–79.

Misell, D.L. and R.E. Burge (1975), Contrast enhancement in biological electron microscopy using two micrographs, J. Microsc. (Oxford) 103, 195–202.

Weszka, J.S., R.N. Nagel and A. Rosenfeld (1974), A threshold selection technique, IEE Trans. Computers C23, 1322–1326.

Chapter 7

Image interpretation

Image analysis and processing can simplify interpretation of, for example, low contrast images and images of multilayered structures. These techniques can also provide more detailed information on the adverse effects of radiation damage and specimen preparation procedures, than is available from direct examination of the original micrograph. In interpreting electron micrographs, even after image processing, there still remain the two fundamental problems of the structural changes occurring during specimen staining (§ 7.1), particularly for the negatively-stained specimens used for studies at high resolution, and during irradiation of the specimen in the electron microscope (§ 7.2). A third problem, the superposition of three-dimensional information in an image (§ 7.3) can be solved by three-dimensional reconstruction from a set of micrographs corresponding to different two-dimensional projections of the specimen (§ 7.4).

7.1 Effects of specimen staining

Unstained specimens of biological material exhibit very low contrast in the electron microscope and the structure is not immediately evident; only if an unstained specimen is highly ordered over an area corresponding to 2000 or more units can image processing produce a sufficiently enhanced image for visual assessment (Unwin and Henderson 1975). Consequently specimens are stained.

Negative staining was originally suggested as a way of imaging isolated viruses with adequate contrast (Brenner and Horne 1959; Horne 1973). Ideally a negative stain is an inert solution which surrounds the molecule, and on drying produces a 'cast' of the original; it is the stain that is imaged, and it is only by implication that the 'stain exclusion pattern' represents the molecule. In addition to the structural changes in the biological material that may result from negative staining, there is very little evidence to suggest that images of negatively-stained specimens give any more information than the overall morphology of the molecules (Aebi et al. 1976, 1977). Whereas a negative-staining solution may completely surround an isolated molecule and hence form a complete outline of it, it may only form a relatively smooth layer on the surfaces of a membrane system or a lipid–protein complex. Most of these solutions are usually composed of large molecular weight units that are unlikely to penetrate into either a lipid bilayer or into small clefts in a large protein. Thus it is likely that the examination of negatively-stained specimens of ordered membrane structures only gives information on the surface topography of the structure. This may also be true for lightly-stained protein crystals. For example, images of tilted crystals of neuraminidase give optical diffraction patterns which are characteristic of a tilted two-dimensional lattice, rather than of a three-dimensional lattice, which would be expected if the negative stain had penetrated into the crystal. There are consequently considerable limitations in the use of the negative-staining technique, although for many specimens there is as yet no useful alternative. At best the negatively-stained specimen represents a low resolution version of the native or unstained material. This can be seen, for example, in a comparison of processed images of stained and 'unstained' (glucose embedded) purple membrane (Fig. 4.1, and Fig. 4.26 in § 4.5), respectively, where the trimer in Fig. 4.1 corresponds to the three protein subunits, each resolved as 7 α-helices in 'unstained' specimens (see also Fig. 7.10 in § 7.4.2).

Furthermore, it cannot be assumed that the negative-staining tecnnique only results in negative staining; it is probable that simultaneous negative and positive staining occur. An excellent example of this is shown by a comparison of images of 'unstained', positively- and negatively-stained catalase crystals using uranyl acetate (Unwin 1975). Figure 7.1 illustrates (a) a negatively-stained thin 'platelet' of catalase, (b) a positively-stained 'platelet' and (c) a negatively-stained single layer; the image of the 'unstained' specimen is not included because it has no visible structure when photographed under subminimal radiation conditions (Unwin and Henderson 1975). The differences between the images are emphasized by an examination of the corresponding optical diffraction patterns (or electron diffraction pattern for the 'unstained' specimen) shown in Fig. 7.2. The electron diffraction pattern obtained from 'unstained' catalase (Fig. 7.2a) should be taken as a reference to be compared with the image transforms of (b) the negatively-

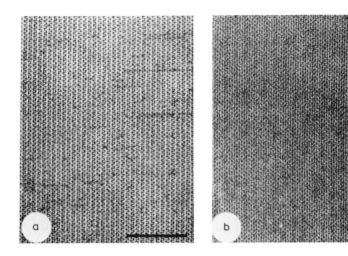

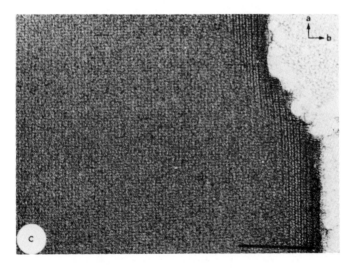

Fig. 7.1. Electron micrographs of (a) a negatively-stained 'platelet', (b) a positively-stained 'platelet' and (c) a negatively-stained single layer of catalase. The stain is uranyl acetate. Image bars = 100 nm. (From Unwin 1975.)

stained 'platelet', (c) the positively-stained 'platelet' and (d) the negatively-stained single layer. The different intensity distributions of the diffraction maxima indicate that the image of the negatively-stained 'platelet' is completely different from the images of either 'unstained' or positively-stained

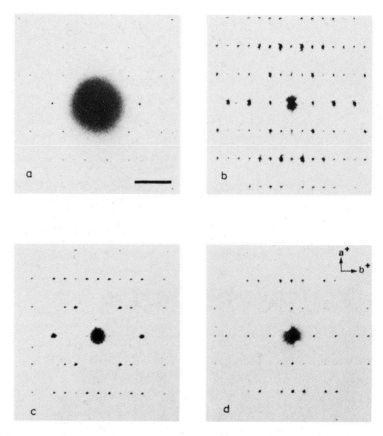

Fig. 7.2.   (a) Electron diffraction pattern from an unstained catalase crystal preserved in glucose and recorded with a low electron dose (less than $10\,e^{-}\,nm^{-2}$ at the specimen); (b–d) optical diffraction patterns from the micrographs shown in Fig. 7.1; (b) a negatively-stained 'platelet', (c) a positively-stained 'platelet', (d) a negatively-stained single layer. Diffraction bar $= 0.2\,nm^{-1}$. (From Unwin 1975.)

catalase. In particular, the alternating strong and weak (or missing) diffraction maxima seen in the horizontal direction of both the electron diffraction pattern of unstained catalase (Fig. 7.2a) and the optical diffraction pattern of the positively-stained specimen (Fig. 7.2c) (characteristic of a $P2_12_12_1$ lattice) are of *similar* intensity in the optical diffraction pattern of the negatively-stained 'platelet' (Fig. 7.2b).

These differences are further emphasized when noise-filtered images (§ 4.2 and § 4.5) are compared at a resolution of 2 nm, the limit set by the staining techniques for catalase. In Fig. 7.3 a comparison is made of the two-

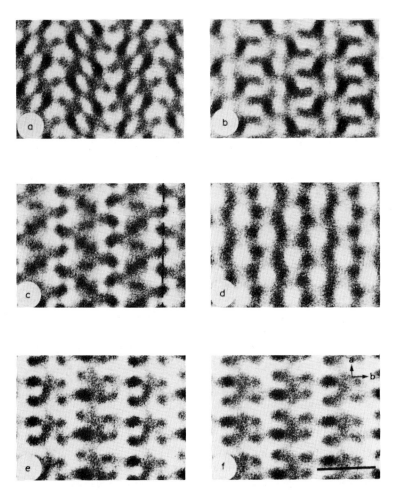

Fig. 7.3 Two nanometre resolution density displays computed from electron micrographs (or from combined micrograph-electron diffraction data (a) and (b), §4.5): (a) a two-dimensional projection of the structure of unstained crystalline catalase for an electron dose ($10\,e^-\,nm^{-2}$) insufficient to damage the protein significantly, (b) this projected structure after it has been 'stabilised' by radiation damage (dose $= 700\,e^-\,nm^{-2}$); (c) and (d) filtered images of negatively-stained and positively-stained catalase 'platelets', respectively; (e) and (f) filtered images of a negatively-stained single layer before (e) and after (f) averaging on the assumption that there is a *b* glide plane. The darker regions represent higher concentrations of more weakly scattering matter: in (a) and (b) this is the protein; in (c), (e) and (f) this is the protein that is not positively-stained with uranyl acetate and in (d) this is any region where little or no positive stain is present. Catalase structure: $P2_12_12_1$. Image bar $= 10\,nm$. (From Unwin 1975.)

dimensional projection of the structure of catalase for (a) an 'unstained' specimen (radiation dose $n_0 = 10\,e^-\,nm^{-2}$ at the specimen), (b) an 'unstained' specimen after radiation damage ($n_0 = 700\,e^-\,nm^{-2}$), (c) and (d) negatively- and positively-stained 'platelets' and (e) a negatively-stained single layer. If the image transforms of the negatively- and positively-stained specimens (Fig. 7.3c and d) are subtracted from each other, the reconstruction that results from this difference transform has a close resemblance to the image of the radiation damaged 'unstained' specimen (Fig. 7.3b). This suggests that the image of catalase negatively-stained with uranyl acetate arises from both negative and positive staining of the protein. This simultaneous positive and negative staining means that some regions of stain indicate the presence of protein and others, its absence, with a consequent uncertainty of interpretation.

Large surface tension forces may occur as the negative-staining solution dries around a large molecule (20–50 nm in diameter) resulting in a severe distortion of the molecule (e.g. Serwer 1977). This is unavoidable for isolated molecules but it does mean that the morphology seen in images should be interpreted with caution. Stabilising an ordered structure against the adverse effects of the negative-staining procedure (e.g. denaturation, dehydration) by first cross-linking with glutaraldehyde or diimido-esters may be useful. For example, Langer et al. (1975) demonstrated that such cross-linking (fixation) of protein crystals preserved the crystal structure to a resolution of 1 nm.

Whenever possible electron microscope observations at high resolution, such as those obtained from negatively-stained specimens, should be verified using other techniques, such as X-ray diffraction on the native material. Although it is usually not possible to prepare single crystals large enough (500 μm) for X-ray crystallography, sufficient information can usually be obtained from either layers of oriented membranes (e.g. Blaurock 1975; Henderson 1975) or small crystals packed together. These specimens give *powder patterns* which are useful in establishing spacings. The scattering of X-rays and electrons by matter are not exactly equivalent, but missing or additional diffraction orders in the optical diffraction pattern of the image of an ordered specimen, as compared with the powder X-ray diffraction pattern of the native material, can indicate false information in the electron image. Other information, such as the molecular weight, conformation and shape of the molecules, obtained by physical and biochemical techniques can also help to confirm the interpretation of electron micrographs (Van Holde 1971).

7.2 Radiation damage

Irradiation with the beam in an electron microscope leads successively to the inactivation (Hahn et al. 1976) and denaturation of the protein molecules (Baumeister et al. 1976a), followed by bond-breakage (Baumeister et al. 1976b) and finally the loss of the three-dimensional structure. It is unlikely that electron micrographs can be taken at the low doses required for the retention of biological activity ($n_0 = 1.5\,e^-\,nm^{-2}$ for catalase), but it is possible to ensure the preservation of the three-dimensional structure ($n_0 = 50$–$100\,e^-\,nm^{-2}$). Table 7.1 lists the radiation doses that cause significant damage in various types of organic and biological structures, and may be used to assess the maximum radiation dose n_0 that can be used in recording high-resolution images (see § 2.2; Table 2.1). The radiation doses listed are averages of those determined in several different studies. Single deter-

TABLE 7.1

Radiation damage to organic and biological specimens as determined from electron diffraction and electron energy loss measurements for 100 keV electrons

Type of structure	Radiation dose		Method[*]
	$e^-\,nm^{-2}$	$C\,m^{-2}$	
Simple structures			
Phthalocyanines	13,000	2000	Diffraction
	150,000	24,000	Energy loss
Aromatic hydrocarbons	6200	1000	Diffraction
DNA bases	1900	300	Diffraction
	23,000	3700	Energy loss
Aromatic amino acids	1100	180	Diffraction
	13,000	2100	Energy loss
Polymers/aliphatic hydrocarbons	380	60	Diffraction
Aliphatic amino acids	120	20	Diffraction
Complex structures			
Negatively-stained proteins	1900	300	Diffraction
DNA/ribosomes	310	50	Diffraction (estimated)
	3800	600	Energy loss
Proteins/membranes	75	12	Diffraction
Hydrated biological specimens	25	4	Diffraction

minations often differ by an order of magnitude (see, for example, the reviews by Stenn and Bahr 1970 and Glaeser 1975). Radiation dose measurements are based either on electron diffraction *decay curves* for crystalline specimens (e.g. Unwin and Henderson 1975) or *electron energy loss measurements* for both crystalline and non-crystalline specimens (e.g. Isaacson et al. 1973). The decay of an electron diffraction pattern detects the destruction of the inter-molecular order in the crystal and is a measure of average damage; electron energy loss measurements give information on the intra-molecular structure of individual macromolecules. Loss of order or crystallinity does not neces-sarily mean that all the individual molecules in a crystal are seriously damaged, but there is no way of distinguishing between damaged and relatively undamaged structural units. Electron energy loss measurements, therefore, give higher radiation dose values for a certain degree of damage of a given structure than electron diffraction data. The results in Table 7.1 refer to an incident electron energy E_0 of 100 keV; increasing E_0 or decreas-ing the temperature of the specimen increases the value of the radiation dose causing significant damage by a factor of up to 4 (Glaeser 1975).

Negatively-stained specimens have intrinsically limited resolution and most of the damage to them appears to occur during the staining process. However, they can be even further damaged by using large radiation doses in the electron microscope. The negative stain can migrate and contract to produce a less accurate representation of the biological structure. This radiation sensitivity of negatively-stained specimens is confirmed by an examination of the optical diffraction patterns of images of catalase crystals, cowpea chlorotic mottle virus arrays, paracrystalline TMV rods and tubulin sheets, negatively-stained with uranyl acetate (T.S. Baker, personal com-munication). With increasing radiation dose there may be a progressive loss of the high order diffraction maxima. The differences between low dose (minimal irradiation conditions, $n_0 = 1000\text{--}2000\ e^-\ nm^{-2}$) and high dose (normal irradiation conditions, $n_0 = 10{,}000\text{--}20{,}000\ e^-\ nm^{-2}$) image pairs are consistent with the conclusion that during irradiation the negative stain migrates prior to crystallisation (Unwin 1974). Images obtained during the initial irradiation period, when the stain is still in an amorphous or glass-like state, indicates that it maps out the fine details of the specimen surface more accurately than after the stain has crystallised (probably in the form of UO_2 for uranyl acetate). Thus images of negatively-stained specimens taken under minimal radiation conditions ($n_0 = 1000\text{--}2000\ e^-\ nm^{-2}$, see below) should provide more reliable structural information than images obtained after normal irradiation ($n_0 > 10{,}000\ e^-\ nm^{-2}$). Preirradiation of the specimen at low magnification (M = 1000–2000), in order to stabilise it against subsequent radiation damage (Dubochet 1973; Johansen 1976), does not completely prevent the loss of the high-order diffraction maxima in the image transform.

Quantitative information on the effect of radiation damage on negatively-stained specimens is limited to crystalline or ordered specimens, because the optical transform of images of isolated particles or molecules is not discrete but continuous, and the specimen transform cannot be distinguished from the noise and carbon support film transforms (§ 3.3). However, it is recommended that minimum irradiation techniques, such as that of Williams and Fisher (1970), are always used when examining molecular structure, since the migration and crystallisation of negative stain, which together produce a poor representation of the morphology of crystalline and ordered specimens, are likely to have even more effect on isolated particles.

Radiation damage of negatively-stained specimens is clearly shown by the example in Fig. 7.4, a processed (averaged) 'one-sided' image of a T-layer from *Bacillus brevis*. The image obtained by using subminimal radiation conditions (n_0 less than 100 e$^-$ nm^{-2}, Fig. 7.4b) shows more detail than even the image taken under minimal radiation conditions (n_0 significantly less than 10,000 e$^-$ nm^{-2}, Fig. 7.4a). Images taken under 'normal' conditions, such as Fig. 7.4c and d, do not show clear evidence of a tetrameric structure

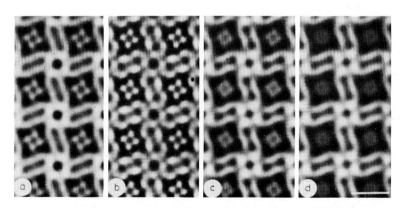

Fig. 7.4. Bright-field images of negatively-stained T-layers from *Bacillus brevis* after image processing, showing the effect of electron dose on structural information. The 'protein' is shown as white on a dark background of negative stain. The electron dose is (a) 5000 e$^-$ nm^{-2} (minimal radiation), (b) 60 e$^-$ nm^{-2} (subminimal radiation), (c) 62,000 e$^-$ nm^{-2} and (d) 620,000 e$^-$ nm^{-2} at the specimen. (b), (c) and (d) were obtained by combining image transform phases and electron diffraction amplitudes (§ 4.5). Image bar = 10 nm. (P.R. Smith and J. Dubochet, unpublished.)

at the centre of the unit cell. The original images used to produce the results in Fig. 7.4 do not give any indication of the adverse effects of radiation damage, so evident in Fig. 7.4 after image processing.

Under standard *minimal radiation* conditions, the specimen is exposed to the electron beam only for a time sufficient to obtain a photographic image with an optical density of 1–2 units ($n_0 \simeq 1000$–$2000\, e^-\, nm^{-2}$). The electron dose can be further reduced by using as low a magnification, M, as possible, because the number of electrons incident on 1 μm^2 of photographic emulsion depends on M^2. It will be remembered (§ 2.2) that the radiation dose on the specimen n_0 ($e^-\, nm^{-2}$ at the specimen) is related to the optical density D on the micrograph taken at magnification M (in units of 1000), thus:

$$n_0 = M^2 D/D_0 \qquad (7.1)$$

where D_0 is the optical density obtained for an electron dose 1 $e^-\, \mu m^{-2}$ on the photographic emulsion (§ 2.2).

The principal problem in taking low magnification (10,000–20,000 ×) images is focusing accurately enough to ensure that the first zero of the phase contrast transfer function $\sin \gamma$ is outside the highest order diffraction maximum observed in the optical transform of an ordered lattice (§ 3.2). Therefore, many pictures are taken so that some are certain to be electron-optically satisfactory.

n_0 can also be decreased by taking micrographs with a lower optical density D. This produces statistically noisy images that can be visually assessed only after image processing. Consequently a significant reduction in D can only be made for images obtained from ordered structures, for which a subsequent averaging over k^2 unit cells can be used to reduce the statistical noise by a factor of k (Kuo and Glaeser 1975; Unwin and Henderson 1975). For aperiodic specimens, or even for specimens showing only rotational symmetry, D cannot be significantly reduced. All specimens should, however, be imaged under minimal radiation conditions using as low an image magnification as possible, compatible with the resolution expected from the specimen. A lower limit for M is set by the resolution of the photographic emulsion ($\simeq 10\, \mu m$); for example, to ensure that the image of a negatively-stained specimen (resolution $\simeq 2\, nm$) is not limited by the photographic emulsion, the minimum value for M should be at least a factor of two greater than 10 $\mu m/2\, nm = 10 \times 10^{-6}/2 \times 10^{-9} = 5000$.

In comparing dark- and bright-field microscopy, one of the most important factors is the higher radiation dose required to obtain a dark-field image with a signal-to-noise ratio comparable to the bright-field image (Table 5.2; § 5.4.1). However, the results of Ottensmeyer et al. (1975a and b, 1977) indicate that small molecules (1–2 nm) may retain their structural integrity even at these high doses.

The use of a scanning transmission electron microscope (STEM) is the most efficient way of producing dark-field images (§ 5.3). The STEM is particularly suited to detecting single heavy atoms used, for example, to label macromolecules (Crewe 1973). It has the additional advantage of effecting a separation of the elastically (E) and inelastically (I) scattered electrons, and can therefore produce the ratio (E/I) or the difference ($E - fI$, where f is an arbitrary constant) image for a thin specimen (§ 5.3.3). The STEM does *not*, however, seem to have a significant advantage over bright-field conventional transmission electron microscopy (CTEM) for the examination of unstained or negatively-stained specimens; the radiation doses necessary to produce an image with a given signal-to-noise ratio in both types of instrument are similar (Misell 1977). For biological sections, which may be relatively insensitive to radiation damage, the main advantage in using the STEM is the absence of the effect of chromatic aberration on the inelastic component, which *is* present in a CTEM image.

7.3   The effect of specimen thickness

For images in which contrast arises mainly from phase contrast, the focus difference $\Delta f = t$ between the top and bottom of a specimen with thickness t provides a fundamental limitation to the interpretation of the image (§ 7.3.1). For specimens imaged under scattering contrast conditions (§ 7.3.2) or for thick specimens, where inelastic scattering predominates, the resolution of the image is limited firstly by the *chromatic aberration* of the objective lens and secondly by the *broadening of the electron beam* in the specimen as a result of *multiple electron scattering* (§ 7.3.3). The fundamental limitation of all two-dimensional images is the superposition of information from a three-dimensional specimen; clearly for a particular structure there will be a maximum specimen thickness, beyond which it is impossible to interpret the two-dimensional projection, even after image processing. If only the surface of the specimen is imaged, as in metal-shadowed specimens or even in some negatively-stained specimens such as T4 phage heads (Aebi et al. 1976, 1977), information is limited to two-dimensions; the fact that the original biological material is a three-dimensional structure does not affect image interpretation, provided no inference is made on the three-dimensional arrangement of the molecules.

7.3.1   High resolution electron microscopy

In the phase object approximation used in the discussion of the theory of image formation in § 3.2, the phase differences between electrons scattered in different elemental slices of the specimen were neglected (Cowley and Moodie 1957); this allows the phase shift $\eta(r)$ to be interpreted directly as the two-dimensional projection of the three-dimensional potential distribution, $V(r, z)$, in the specimen. The maximum phase shift, $\Delta(v) = \pi\lambda tv^2$, is equivalent to a focus difference t between the top and bottom of the specimen; that is, $\Delta f = t$ in the defocus term of γ in Eq. (3.19) in § 3.2. Although this phase difference does not adversely affect the overall resolution of the image, it does place a limit on it, beyond which the image can no longer be interpreted as the projection of $V(r, z)$ (Misell 1976). This interpretation will only be valid if the phase difference $\Delta(v)$ is significantly less than one. If $\Delta(v)$ is chosen to be less than 0.2 radians, the resolution $r\,(= 1/v)$, to which an image can be correctly interpreted as a projection of $V(r, z)$, is (Lipson and Lipson 1972):

$$r \simeq 4.0 \sqrt{\lambda t} \tag{7.2}$$

Thus for $\lambda = 3.7$ pm ($E_0 = 100$ keV), the resolution limits are 0.8 nm for $t = 10$ nm, and 1.7 nm for $t = 50$ nm. Although detail smaller than these values is present in the reconstructed image it *cannot* be readily interpreted.

To obtain quantitative information on the effect of the phase difference $\Delta(v)$, the image transform $J(v)$ is examined (§ 3.2). $\Delta(v)$ can be considered as a phase error, which is added to $\gamma(v)$ in the phase contrast transfer function $\sin \gamma(v)$, so that the image transform of a weak phase object becomes (Eq. (3.24) in § 3.2):

$$J(v) = \delta(v) - 2A(v) \sin \left[\gamma(v) + \Delta(v)\right] \tag{7.3}$$

where $A(v)$ is the Fourier transform of the phase shift $\eta(v)$.

The mean value of the phase error, Δ, will in fact only be $\pi\lambda tv^2/2$, the phase difference between the centre of the specimen and the top or bottom. Figure 7.5 shows the variation of the mean value of Δ with spatial frequency $v\,(= 1/r)$. Because Δ depends on v^2, the phase error is much more important at high

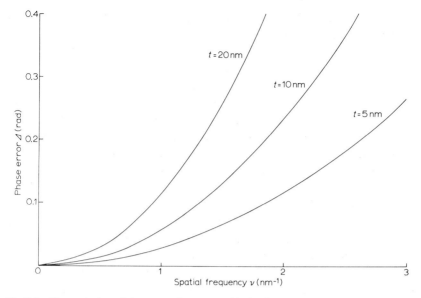

Fig. 7.5. The variation of the mean phase error $\Delta(v)$ (radians) in the image transform with spatial frequency v (nm^{-1}) for specimens of thickness $t = 5$ nm, 10 nm and 20 nm. (From Misell 1976.)

resolution ($v = 2$ nm^{-1}) than at lower resolution ($v = 0.5$ nm^{-1}); Δ depends linearly on the specimen thickness, t. The effect of Δ on the image transform will be equivalent to an error in determining γ from the optical transform (§ 3.3.4). If the approximation is made that Δ is so small that $\sin \Delta \simeq \Delta$ and $\cos \Delta \simeq 1$, the $\sin (\gamma + \Delta)$ term in Eq. (7.3) can be expanded to give:

$$J(v) = \delta(v) - 2A(v) \{ \sin [\gamma(v)] + \Delta(v) \cos [\gamma(v)] \} \qquad (7.4)$$

Thus the effect of specimen thickness (or an equivalent defocus error) is to add a term $2A\Delta \cos \gamma$ to the image transform of a weak phase object; this term is then a measure of the error in the image transform as a result of imaging a specimen of thickness t. It is suggested that the high frequency components of the image transform, which are those most seriously affected by this error term, are reduced by a factor depending on the relative magnitudes of $\sin \gamma$ and $\Delta \cos \gamma$ in Eq. (7.4). Although the effect of the error term, $\Delta \cos \gamma$, on the image transform cannot be corrected for, because only its average value is known, it is reasonable to reduce the contributions from high frequency Fourier components of the image transform, J, so that in the final two-dimensional map of $\eta(r)$, high resolution information is suitably weighted so as to reflect its veracity to the original structure, a projection of $V(r, z)$.

7.3.2   Phase contrast versus scattering contrast

Phase contrast images are obtained whenever the image defocus Δf is chosen to maximise $\sin \gamma$. This corresponds to an underfocus of $r^2/2\lambda$ for a specimen spacing r (§ 3.2.2). Thus phase contrast images can be obtained from ribosome crystals with a lattice spacing of 59.5 nm by using underfocus values as large as 478 μm for the (0, 1) diffraction maximum (§ 3.4.3). Such large underfocus values produce an image with a phase contrast transfer function $\sin \gamma$ that oscillates rapidly at larger spatial frequencies, and the image is difficult to interpret reliably except for spacings $r \simeq 59.5$ nm. The only way to use phase contrast images taken at such large underfocus values is to combine several image transforms from a focus series, and to apply a correction for the effects of $\sin \gamma$ (§ 4.4 and § 4.5).

Normally it is recommended that images of low resolution specimens (e.g. biological sections) or images of specimens with large spacings (e.g. ribosome crystals) are taken just underfocus, so that image contrast arises mainly from scattering contrast; the image intensity is then related to mass thickness variations ρt in the specimen (§ 5.1). If the specimen thickness is constant, the optical density variations in the image are effectively a measure of the density variations ρ across the specimen. Note that a phase contrast image is interpreted in terms of the phase shift $\eta(r)$, proportional to the projection of the potential distribution $V(r, z)$ in the specimen, subject to the limitations discussed in § 7.3.1; a scattering contrast image, however, is interpreted in terms of density variations $\rho(r)$ in the specimen (the *logarithm* of the image intensity is proportional to ρ). In practice this distinction may not be important, but it is essential not to process a scattering contrast image as if it were a phase contrast image; for example, applying a correction for $\sin \gamma$ to a scattering contrast image will produce a false reconstruction, because the image transform does not include a significant contribution from $A \sin \gamma$.

As the specimen thickness increases, a larger proportion of the incident electron beam will be inelastically scattered. It is particularly important to be aware of this when using unstained specimens or specimens composed of low atomic number (Z) elements, since the ratio of elastic scattering to inelastic scattering in the specimen decreases with decreasing Z. In an unstained section of thickness 100 nm, approximately 72% of the incident electron beam will be inelastically scattered and only 15% elastically scattered. Since inelastically scattered electrons are out of phase with each other, that is they are *incoherent*, as a result of their large energy spread, they do not give any significant phase contrast effects. Images with a significant contribution from inelastically scattered electrons are essentially scattering contrast images (Burge 1973).

Another effect of increasing specimen thickness is that the average angle of scattering θ increases, and the resolution of the image is no longer limited by spherical aberration but by the broadening of the electron beam in the specimen (Reimer and Gentsch 1975), and, because the image is dominated by inelastic scattering, chromatic aberration may be the main factor limiting resolution (Cosslett 1956). In the next section the resolution limitations resulting from these two factors, *beam broadening* and *chromatic aberration*, in scattering contrast images are evaluated.

7.3.3 Resolution limitations

Electrons are scattered over a range of angles θ (Fig. 2.7; § 2.5.2) and, for a particular angle of scattering θ, the broadening, d, of the electron beam at the exit face of a specimen of thickness t is:

$$d = 2\theta t \tag{7.5}$$

As the specimen thickness increases the average value of θ increases as a result of multiple scattering, and for a thick specimen the broadening of the electron beam d may limit image resolution. To calculate a *single* value for d it is necessary to choose a value of θ that is a measure of the angular broadening of the electron beam. The value of θ chosen corresponds to $\alpha_{1/2}$ (radians), the semi-angle which contains 50% of the scattered electrons (elastically or inelastically scattered). This gives an arbitrary definition for the image resolution $d_{1/2} = 2\alpha_{1/2}t$, as limited by beam broadening, and the actual profile of the beam should be calculated using the actual angular distribution (e.g. see Fig. 2.7; § 2.5.2). The mean value of θ could equally well have been chosen, but this does give a value for d that is biased towards the larger angles of scattering. Figure 7.6 shows the variation of $\alpha_{1/2}$ with specimen thickness for carbon ($Z = 6$), aluminium ($Z = 13$) and gold ($Z = 79$). The value of $\alpha_{1/2}$ for inelastic scattering is smaller than the value of $\alpha_{1/2}$ for elastic scattering, but the differences become smaller for thick specimens, where the inelastic

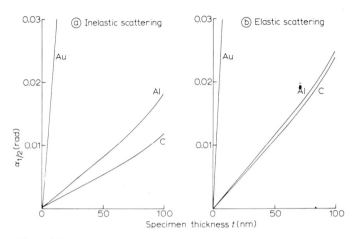

Fig. 7.6.   The variation of the semi-angle, $\alpha_{1/2}$ (radians), that includes 50% of (a) the inelastically and (b) the elastically scattered electrons for carbon ($Z = 6$), aluminium ($Z = 13$) and gold ($Z = 79$) with increasing specimen thickness t. Incident electron energy $E_0 = 100\,\text{keV}$.

angular distribution is broadened by further elastic scattering. Note that these electrons are still classified as inelastically scattered, even though they may have been additionally elastically scattered. Although the angular distributions of elastically and inelastically scattered electrons are similar for thick specimens, the essential distinction between the two is that the elastic component has not lost any energy in the specimen, whereas the inelastic component has an energy spread, and is therefore affected by chromatic aberration. It can be seen from Fig. 7.6 that the value of $\alpha_{1/2}$ for gold increases very rapidly with thickness, indicating that beam broadening will limit the resolution for stained specimens at a smaller thickness than for unstained specimens.

Figure 7.7 shows how $d_{1/2} = 2\alpha_{1/2}t$ increases with specimen thickness for carbon (density $\rho_c = 2000\,\text{kg m}^{-3}$). For a thickness less than 50 nm, beam broadening for both the elastic and inelastic components is less than 0.5 nm, and is comparable to the resolution limit set by the spherical aberration of the objective lens. However, $d_{1/2}$ increases proportionally to t^2, so that for an unstained specimen of thickness 100 nm, the elastic image resolution is significantly larger than 2 nm. Stained specimens will reach this resolution limit sooner. Note that the thickness scale in Fig. 7.7 refers to carbon, but an *equivalent* thickness of carbon for a biological section of density ρ_s can be calculated by multiplying the specimen thickness t by the ratio ρ_s/ρ_c. For example, an unstained section of thickness 100 nm and density $1300\,\text{kg m}^{-3}$

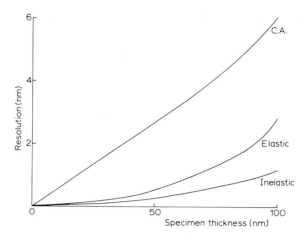

Fig. 7.7. The variation of the image resolution, as limited by beam broadening, $d_{1/2} = 2\alpha_{1/2}t$, for inelastic and elastic scattering with increasing thickness of carbon, t (nm). The resolution limit resulting from chromatic aberration (C.A.) in the inelastic image is also shown. $E_0 = 100$ keV, $C_c = 2$ mm.

is approximately equivalent to $100 \times 1300/2000 = 65$ nm of carbon. An equivalent thickness of carbon can also be used for stained sections, provided that the density of the section is not much greater than 2000 kg m^{-3}.

The inelastic image resolution is not, however, limited by beam broadening, but by chromatic aberration. The chromatic aberration of the inelastic image is calculated from:

$$r_c = C_c \theta \Delta E / E_0 \qquad (7.6)$$

where $\theta = \alpha_{1/2}$ and ΔE is the most probable energy loss, which is about 25 eV for biological materials (§ 2.5.2). The variation of chromatic aberration with specimen thickness is shown in Fig. 7.7 by the curve C.A.. Clearly the chromatic defect determines the resolution of the inelastic image. These results are slightly lower than estimates of chromatic aberration based on the mean energy loss (Cosslett 1956), but about the same as results based on the energy half-width of the energy loss distribution of the inelastically scattered electrons (Cosslett 1969).

A scattering contrast image is a superposition of an elastic image, whose resolution is limited by beam broadening, and an inelastic image with a resolution determined by chromatic aberration. Whether the image resolution is limited by beam broadening or by chromatic aberration depends on the relative proportions of elastic and inelastic scattering contributing to the image. For thin biological sections (20–50 nm) chromatic aberration does not limit image resolution, because a significant proportion of the higher resolution elastic component contributes to the image, and the inelastic scattering merely forms an undesirable low resolution background, reducing image contrast. However, for thick sections (> 50 nm) the elastic scattering contributing to the image is small compared to the inelastic scattering, so that the appearance of the image is dominated by the contribution from the inelastic component, and its resolution is limited by chromatic aberration.

Decreasing the incident electron energy, E_0, increases the interaction of the electron beam with the specimen, so that the value of $\alpha_{1/2}$ increases as E_0 is decreased. Thus both beam broadening and chromatic aberration increase at low incident electron energies (20–60 keV); the increase in chromatic aberration is more marked than beam broadening, because the factor $\Delta E/E_0$ in Eq. (7.6) significantly increases at low incident electron energies. For optimum resolution in a scattering contrast image of a biological section, a high incident electron energy (100 keV) should, therefore, be used. However, increasing E_0 decreases the mass scattering cross-section, S_p, and the scattering contrast is consequently reduced (§ 5.1). Imaging biological sections becomes a compromise between resolution and contrast; the use of an incident electron energy of 100 keV is recommended provided that the image contrast is high enough for a visual assessment of the image. It may also be possible to enhance the image contrast by spatial filtering (§ 4.7) or thresholding (§ 6.3), in addition to the use of a small objective aperture (§ 5.1).

It may be advantageous to examine thick specimens in the STEM because, although the elastic and inelastic scattering cannot be separated on the basis of the differences in their angular distributions, the inelastic component is unaffected by chromatic aberration. Thus the resolution of the image of a thick specimen in the STEM will be limited only by beam broadening (Gentsch et al. 1974).

7.4   Limitations of two-dimensional information

The information in an image is restricted either to the surface morphology of the specimen or to a two-dimensional projection of its three-dimensional structure. The arrangement and conformation of the subunits making up the structure cannot, in general, be inferred from the image, and since biological function is closely related to the three-dimensional structure and the arrangement of molecules, two-dimensional projections are of limited use at the molecular level. Sometimes by combining image information with information obtained by other techniques, the three-dimensional structure can be inferred. For example, X-ray diffraction from the purple membrane of *Halobacterium halobium* shows that the protein in the membrane exists mainly in an α-helical conformation (Henderson 1975); thus in the projected structure of the purple membrane the dense regions of protein in the contour map (Fig. 4.26; § 4.5) were interpreted as α-helices arranged perpendicular to the plane of the membrane (Unwin and Henderson 1975). This arrangement of the protein in the purple membrane was confirmed by a three-dimensional reconstruction from images of tilted specimens (Fig. 7.10 in § 7.4.2).

A normal image represents the two-dimensional projection of the three-dimensional potential distribution (for phase contrast images) or density distribution (for scattering contrast images) in the specimen. If several two-dimensional projections are obtained, corresponding to different views of the structure, it is possible to combine them to produce a three-dimensional model of the specimen. A microscope with a good, eucentric tilting stage is required. Three-dimensional reconstruction can be achieved only by numerical methods: there are no optical methods for combining two-dimensional projections into a three-dimensional structure. The synthesis of a three-dimensional structure from its two-dimensional projections can either be made in real space (Frank 1973; Vainshtein 1973) or in Fourier space (DeRosier and Klug 1968; Crowther et al. 1970a; DeRosier 1971; Frank 1973). In Fourier space, the transform from each two-dimensional projection contributes a *central-section* to the three-dimensional Fourier transform of the three-dimensional structure (§ 7.4.1), and thus by combining these central-sections a three-dimensional transform is produced. The inverse Fourier transform then gives a reconstruction of the three-dimensional structure. Fourier techniques are preferred to real space techniques, because an objective assessment can be made of the resolution of the reconstruction and of the effect on the reconstruction of using only a limited range of tilt angles (Crowther et al. 1970a; Klug and Crowther 1972). In practice the maximum tilt angle obtainable in most goniometer or tilt stages is $\pm 60°$, where the theoretical requirement of specimen tilt angles is $\pm 90°$. The problem of increased radiation damage as a result of taking several images of the same specimen area can be avoided with a crystalline or ordered specimen. Tilted images of the specimen can be recorded from different, previously unirradiated, areas (Henderson and Unwin 1975). In addition, the symmetry of the lattice can be used to reduce the number of two-dimensional projections required (§ 7.4.2). For a helical particle only a single two-dimensional projection is required to produce a three-dimensional reconstruction (§ 7.4.3; DeRosier and Klug 1968; see also § 3.6). For isolated particles the different projections may be obtained from a single particle by using a tilt stage or from several particles in different but identifiable orientations. The latter procedure is only practical if the particle has a high degree of symmetry, such as spherical symmetry (Crowther et al. 1970b, 1975). For asymmetric particles or particles with low symmetry, such as the hexon from adenovirus, it is very difficult to relate the different views seen in the micrograph (see Fig. 4.29; § 4.7.1), without some additional information obtained from images of the tilted specimen (§ 7.4.1).

7.4.1   Three-dimensional reconstruction: principles

Three-dimensional reconstruction using Fourier techniques is based on the *central-section* or *projection theorem*: the Fourier transform of a two-dimensional projection of a three-dimensional object is a *central-section* or plane of the three-dimensional Fourier transform of the object. The origin of each central-section corresponds to the origin $(0, 0, 0)$ of the three-dimensional transform provided that the projections are aligned so as to have a common origin; that is, the two-dimensional transforms have the same *phase origin* (Crowther et al. 1970a). In order to obtain the three-dimensional transform of the object, N projections are required, each separated by a tilt angle of θ. These N views are used to calculate N central-sections, each separated by an angle θ, in the three-dimensional transform. Figure 7.8 shows a plan view of two such central-sections; the angular separation θ, of the two sections has been exaggerated so the figure is not to scale. The frequency sampling on each plane is $\Delta v \simeq 1/D$, where D is the diameter of the particle. As the distance from the centre of the three-dimensional transform increases, the separation of the planes increases, so that the sample points in the three-dimensional transform are closely spaced at low spatial frequencies (low resolution) but widely spaced at high spatial frequencies (high resolution). Ideally the projections should be closely enough spaced so that in the three-dimensional transform the maximum distance between sample points on two adjacent central-sections is less than, or equal to, Δv; that is, the same as the frequency sampling on each central-section. At some maximum frequency v_{max}, corresponding to a resolution $d = 1/v_{max}$ in the reconstruction of the object, the separation between adjacent central-sections should be less than Δv. Thus from the geometry of Fig. 7.8, the angle θ between projections should be chosen so that:

$$\Delta v > v_{max}\theta \qquad (7.7)$$

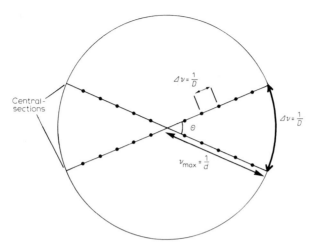

Fig. 7.8. The calculation of the number of projections, N, required to make a three-dimensional reconstruction of an object, size D, at a resolution d. The numerical transform is sampled at approximately $\Delta v = 1/D$ on each plane (central-section), and successive planes are rotated by $\theta = \pi/N$. The maximum spatial frequency v_{max} determines the resolution of the three-dimensional reconstruction, $d = 1/v_{max}$.

it is assumed that there are N equally-spaced projections so that $\theta = 180°/N = \pi/N$ radians. The resolution, $d = 1/v_{max}$, of the reconstruction made from the three-dimensional transform is determined by replacing Δv by $1/D$ and θ by π/N in Eq. (7.7):

$$d > \pi D/N \qquad (7.8)$$

Normally the inequality is replaced by equality, so that the resolution, d, of a three-dimensional reconstruction of an object of diameter D made from N equally-spaced projections, is:

$$d = \pi D/N \qquad (7.9)$$

Note that in Eq. (7.9) it is assumed that projections are obtained for all tilt angles between $-90°$ and $+90°$. The practical limit of $\pm 60°$ means there will be a region (33%) of the three-dimensional transform with no sample points; thus the three-dimensional reconstruction will be *anisotropic*, unless this missing section of the Fourier transform can be generated by the symmetry of the object. Also it has been assumed that the particle is characterised by a single diameter D. A solid three-dimensional object has three dimensions D_x, D_y and D_z, so that the resolution of the reconstruction in each of these three orthogonal directions will be determined from Eq. (7.9) by using the respective values D_x, D_y and D_z. Usually, however, D is chosen to be the largest dimension of the object so that the resolution calculated from Eq. (7.9) provides an upper limit to the resolution of the reconstruction.

For a particle with no symmetry, the minimum number of views required is large; for example, a ribosome of diameter $D \simeq 25$ nm requires 39 independent views for a three-dimensional reconstruction at a resolution of 2 nm. ($N = \pi D/d = \pi \cdot 25/2 \simeq 39$). However, if the particle has m-fold symmetry, the number of views required for a given resolution is reduced by a factor of m. For example, the spherical human wart virus has a diameter, D, of 56 nm, which for a reconstruction at 6 nm resolution requires 29 projections; but the subunits of the virus are arranged on an icosahedral lattice with 3-fold and 5-fold axes of symmetry. Thus, provided two views of the virus are selected that correspond to projections along the 5- and 3-fold axes, only $29/(5 \times 3) \simeq 2$ projections are required for a reconstruction at 6 nm resolution (Crowther et al. 1970b).

Often it will not be possible to obtain a sufficient number of views for a three-dimensional reconstruction at a resolution d; the missing sections of the three-dimensional transform are then calculated by interpolation using the values of the three-dimensional transform that are known. Several different interpolation methods, including the simple bi-linear interpolation (Eq. (4.19); § 4.2.3, extended to three dimensions), can be used to produce a uniformly sampled three-dimensional transform (Crowther et al. 1970a; Smith et al. 1973). It is also usual to interpolate the three-dimensional transform from a polar coordinate system (Fig. 7.8) to Cartesian coordinates, so that sample points are equally spaced in the X, Y, and Z directions of the three-dimensional transform, $F(X, Y, Z)$. This allows the inverse transform of $F(X, Y, Z)$, $\rho(x, y, z)$, to be calculated using a three-dimensional fast Fourier transform (Ten Eyck 1973); the use of the FFT is essential for the large transform arrays that are normally generated in a three-dimensional reconstruction. It is possible to produce a three-dimensional reconstruction with an *apparent* resolution, d, even if the number of views, N, is much less than specified by Eq. (7.9). The three-dimensional transform can always be interpolated out to a spatial frequency $v_{\max}$ ($= 1/d$) with a sampling $\Delta v = 1/D$; but, because of the small number of central-sections used in the interpolation, the interpolated values of the transform will be incorrect, particularly at the higher spatial frequencies. Thus, when an inverse transform is calculated, any fine detail in the object is a result of using incorrect transform values.

The maximum value of v, v_{max}, to which the transform can be reliably interpolated should always be determined from Eq. (7.9) using the actual number (N) of projections available, including additional projections that can be generated by the symmetry of the object; that is, $v_{max} \simeq N/\pi D$.

7.4.2 Three-dimensional reconstruction: crystalline specimens

Radiation damage can be minimised by recording images of the tilted specimen from different, previously unirradiated, areas, provided that the crystal or ordered lattice has a single orientation on the grid. If the crystals have several orientations on the grid it may be possible to relate the different projections using *crystallographic symmetry* (Ohlendorf et al. 1975). Because crystalline specimens usually extend over 100–500 nm, there is a large *focus gradient* across the image, perpendicular to the axis of tilt, for large angles of tilt. For example, in the determination of the three-dimensional structure of the purple membrane (Henderson and Unwin 1975) the specimen extended over about 100 unit cells, so that for the largest angle of tilt (57°), the focus difference between extremes of the specimen was 400 nm. Thus parallel to the axis of tilt, the image had a constant focus, but perpendicular to the axis of tilt, a correction had to be applied to the image for the change in sin γ as a result of the focus gradient (Henderson and Unwin 1975). Since the purple membrane is only a single unit cell thick ($c = 4.5$ nm), the three-dimensional transform of the membrane is continuous in the c^* direction (perpendicular to the plane of the membrane) but discrete in the a^* and b^* directions (parallel to the plane of the membrane), where the lattice extends over a large number of unit cells (§ 3.1). Central-sections of the three-dimensional transform were obtained by taking images at particular tilt angles (Fig. 7.9), using specially designed specimen holders. The three-dimensional transform was calculated using the phases derived from the image transforms, after correction for the effect of sin γ, and the amplitudes were determined from the corresponding electron diffraction patterns (§ 4.5: Henderson and Unwin 1975). Henderson and Unwin used 18 images taken at tilt angles between 0° and 57°. Because of the symmetry of the lattice, this gives 36 crystallographically independent lattice planes or central-sections in the three-dimensional transform. Partly as a result of the limited angle of tilt (57°), the three-dimensional reconstruction of a single protein unit of the purple membrane (Fig. 7.10) has a resolution of 0.7 nm in the plane of the membrane but only 1.4 nm perpendicular to the plane of the membrane. This anisotropic resolution in the three-dimensional model is also a result of the low

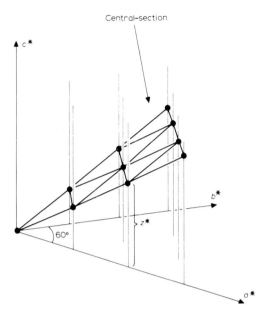

Fig. 7.9.   Part of the three-dimensional reciprocal lattice showing the geometry of the lattice lines in the hexagonal space group P3. a^*, b^* and c^* are the reciprocal lattice vectors. a^* and b^* lie in, and c^* is perpendicular to the plane of the membrane. A central-section, which is perpendicular to the incident electron beam, has been drawn through the lattice. The intersection of this central-section with the reciprocal lattice is determined by the angle of tilt and the axis about which the membrane is tilted. Individual diffraction patterns and micrographs provide the amplitudes and phases in the section at the points shown. z^* represents the coordinate along the c^* direction of one of these points. The accuracy of measurement of both the amplitudes and phases depended on having sharp lattice lines. It is therefore necessary to ensure that, on the microscope grid, the membranes remained coherently ordered and flat to within $1/5°$. (From Henderson and Unwin 1975.)

intensities of the diffraction maxima at z^* values greater than $0.8\ nm^{-1}$ in the c^* direction (Fig. 7.9). The three-dimensional structure (Fig. 7.10) is dominated by rod-shaped features aligned perpendicularly to the plane of the membrane. There are 7 rods in each *asymmetric unit* (single protein unit) of the crystal (International Tables for X-ray Crystallography, Vol. I), and these are packed 1–1.2 nm apart. X-ray diffraction studies indicate that these rods are α-helices (Blaurock 1975; Henderson 1975).

7.4.3   *Three-dimensional reconstruction: helical particles*

The three-dimensional arrangement of the protein subunits on a helix can be determined from a single projection, since the micrograph of a helical

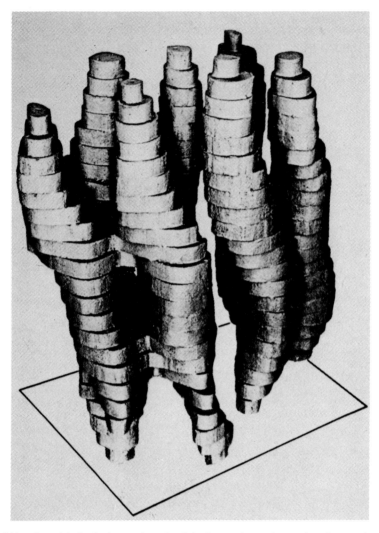

Fig. 7.10.   A model of a single protein molecule in the purple membrane viewed approximately parallel to the plane of the membrane. The most strongly tilted α-helices are in the foreground. Note that the complete protein unit in the membrane consists of three of these molecules arranged on a P3 lattice. (From Henderson and Unwin 1975.)

structure provides several views of each protein subunit (DeRosier and Klug 1968). The resolution of the reconstruction is limited by the number of structural units in the repeat distance of the helix c (§ 3.6). The first step in a three-dimensional reconstruction is to extract from the image transform

those parts that are characteristic of the helix, that is, the layer line amplitudes *and* phases (§ 3.6.3 and § 3.7.3). These layer line terms, $G_n(R, Z)$,
correspond to the Fourier coefficients in the Fourier expansion of the three-
dimensional transform, $F(R, \Phi, Z)$, of the helix $\rho(r, \phi, z)$ in terms of its *angular
harmonics n* (Klug et al. 1958; DeRosier and Moore 1970); that is:

$$F(R, \Phi, Z) = \sum_n G_n(R, Z) \exp \left[in(\Phi + \pi/2) \right] \qquad (7.10)$$

The three-dimensional structure of the helix, $\rho(r, \phi, z)$, can then be calculated
by taking an inverse transform of F. Because of the difficulty in extracting
layer lines from a noisy image transform, it is usual to add together the
transforms from several helical particles so as to increase the signal-to-noise
ratio of the Fourier coefficients G_n (Amos and Klug 1975; Smith et al. 1976).
Figure 7.11a is the two-dimensional projection of the helix from a T4
bacteriophage tail, obtained after the addition of the image transforms of
several aligned helical particles (Smith et al. 1976). The z-sections (Fig. 7.12),

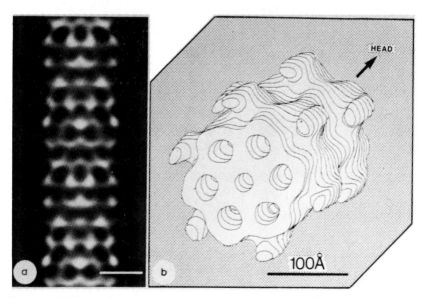

Fig. 7.11.   (a) The projected image obtained by averaging several helically filtered and mutually
aligned T4 phage tails. It corresponds to what would have been obtained if the averaged three-
dimensional reconstruction were reprojected. The helical repeat is (7×4.1) nm. (b) shows a
serial-section plot of two axial repeats of the reconstruction, contoured at a particular level.
The contour is taken to divide the reconstruction sharply into two regions, one of which is
taken to be 'protein' (absence of stain) and the other the stain surrounding the 'protein'. (From
Smith et al. 1976.)

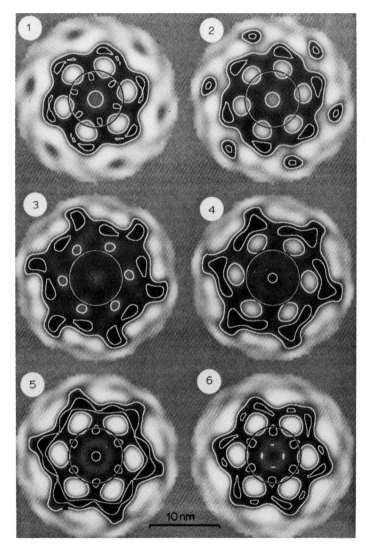

Fig. 7.12. Six equally-spaced z-sections, 0.68 nm apart, taken through the averaged recon-struction of one axially repeating unit ($p = 4.1$ nm) of the T4 phage tail. Section 1 corresponds to the first section seen in the three-dimensional reconstruction (Fig. 7.11b). The contour lower level shown here is the level at which Fig. 7.11b was contoured. The circle in the centre of each section corresponds to an estimate of the mean diameter of the tail tube; the position of the actual boundary between the tube and sheath is not known. The sections are viewed from the baseplate looking towards the head. 'Protein' is black and the background grey is the zero level of the reconstruction. (From Smith et al. 1976.)

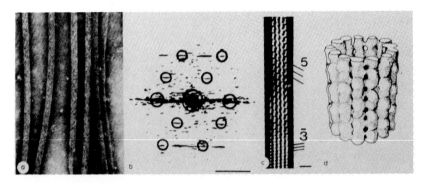

Fig. 7.13. (a) An electron micrograph showing negatively-stained microtubules at the tip of a flagellum from *Trichonympha* (image bar = 100 nm); (b) an optical diffraction pattern from a straight section of microtubule (diffraction bar = 0.2 nm^{-1}); (c) the filtered image (image bar = 20 nm) produced using the mask drawn on the diffraction pattern, which includes contributions from only one side of the microtubule. In the lower part of the image (c) the 8 nm periodicity is very weak and the 4 nm helices (marked $\bar{3}$) can easily be followed. In the rest of the image (c) the 8 nm periodicity dominates giving rise to dumbell-shaped units. This result is in agreement with the three-dimensional reconstruction (d), showing a 8 nm dimer composed of two components α and β approximately 4 nm in diameter. The 8 nm helices of dimers are marked by the number 5 in (c), which is the index for this family of helices. (From Amos and Klug 1974; and L.A. Amos, unpublished.)

corresponding to vertical sections through the helix, were combined in a *serial-section* plot to give a three-dimensional perspective of the T4 tail (Fig. 7.11b; Smith et al. 1976).

Microtubules from flagella and cilia also consist of protein subunits arranged on a helix; here the situation is complicated by the fact that there are two sets of helices with different periodicities (Amos and Klug 1974). A one-sided reconstruction (Fig. 7.13c) using the optical diffraction pattern (Fig. 7.13b) from the image of a single microtubule (Fig. 7.13a) is not very informative (§ 4.3). However, the three-dimensional reconstruction in Fig. 7.13d shows clearly how the protein subunits, α and β, are arranged on the surface of the microtubule (Amos and Klug 1974).

7.4.4   Three-dimensional reconstruction: asymmetric particles

The number of different projections, N, required for the three-dimensional reconstruction of an asymmetric molecule at a resolution d cannot be reduced significantly below the number calculated using Eq. (7.9). For even a small particle, such as the hexon from adenovirus ($D \simeq 8$ nm), at least 10 projections are required for a three-dimensional reconstruction at 2 nm resolution.

There are two alternative methods for determining the three-dimensional structure of an asymmetric molecule. Either a large number of projections of a single particle is determined or a three-dimensional model is built, based on the various projections of the particle seen in a single micrograph (e.g. Ottensmeyer et al. 1975a and b).

In any series of micrographs taken of a single particle, there is a progressive increase in the radiation dose given to the specimen. The accuracy of a three-dimensional reconstruction is, therefore, unlikely to be very good. When using either real space or Fourier techniques, the two-dimensional projections must be first aligned, so that they have a common origin. Using Fourier methods the phase origin of each central-section must be determined, before an inverse three-dimensional Fourier transform can be calculated (Crowther et al. 1970a); this is very difficult to achieve for particles that have little or no symmetry (§ 3.3.2).

A model of the molecule can be constructed using, for example, plasticine or polystyrene. For modelling unstained molecules this model can be X-rayed and *transmission radiographs* produced for different orientations of the model (Caspar 1966; Ottensmeyer et al. 1975a). The radiographs can then be compared with micrographs, which include molecules in various orientations. It is not sufficient to illuminate the model with light and examine the shadows cast by the model at various angles, because the views of the molecule seen in a micrograph correspond to projections *through* the molecule. When modelling negatively-stained molecules it is essential to model the negative stain profile around the molecule (Barrett et al. 1977), because it is the negative stain that is imaged and *not* the molecule (Caspar 1966; Horne 1973). One problem in modelling negatively-stained molecules is that the stain profiles seen in a micrograph do not represent projections of the same three-dimensional structure, unless the molecule is completely embedded in the stain (Horne 1973). Model building does not eliminate the need for images of tilted specimens, because the model should always be *correlated* (§ 4.7.2) with projections seen in the micrograph, that have a known relationship to each other (e.g. their relative orientations on the specimen grid). The most flexible method of model building is to construct a three-dimensional model using a computer with a television or graphics display. This computer model can then be rotated through specified angles, and compared with particular projections seen on the micrograph or a suitably enlarged print (e.g. Fig. 4.29; § 4.7.1). The model can be altered so as to improve the agreement with all or most of the projections seen in the image (Barrett et al. 1977).

References

Aebi, U., R.K.L. Bijlenga, B. ten Heggeler, J. Kistler, A.C. Steven and P.R. Smith (1976), A comparison of the structural and chemical composition of giant T-even phage heads, J. Supramol. Struct. *5*, 475–495.

Aebi, U., R. van Driel, R.K.L. Bijlenga, B. ten Heggeler, R. van den Broek, A.C. Steven and P.R. Smith (1977), Capsid fine structure of T-even bacteriophage: binding and location of two dispensible capsid proteins into the P23* surface lattice, J. Mol. Biol. *110*, 687–698.

Amos, L.A. and A. Klug (1974), Arrangement of subunits in flagellar microtubules, J. Cell Sci. *14*, 523–549.

Amos, L.A. and A. Klug (1975), Three-dimensional image reconstruction of the contractile tail of T4 bacteriophage, J. Mol. Biol. *99*, 51–73.

Barrett, A.N., R.K. Chillingworth, W.J. Perkins and N.G. Wrigley (1977), Simulation of single molecule images combining computer and electron micrograph analysis, in: Developments in electron microscopy and analysis, D.L. Misell, ed. (Institute of Physics, Bristol and London), pp. 131–134.

Baumeister, W., U.P. Fringeli, M. Hahn and J. Seredynski (1976a), Radiation damage of proteins in the solid state: changes of β-lactoglobulin secondary structure, Biochim. Biophys. Acta *453*, 289–292.

Baumeister, W., M. Hahn and J. Seredynski (1976b), Radiation damage of proteins in the solid state: changes of amino acid composition in catalase, Ultramicroscopy *1*, 377–382.

Blaurock, A.E. (1975), Bacteriorhodopsin: a trans-membrane pump containing α-helix, J. Mol. Biol. *93*, 139–158.

Brenner, S. and R.W. Horne (1959), A negative staining method for high resolution electron microscopy of viruses, Biochim. Biophys. Acta *34*, 103–110.

Burge, R.E. (1973), Mechanisms of contrast and image formation of biological specimens in the transmission electron microscope, J. Microsc. (Oxford) *98*, 251–285.

Caspar, D.L.D. (1966), An analogue for negative staining, J. Mol. Biol. *15*, 365–371.

Cosslett, V.E. (1956), Specimen thickness and image resolution in electron microscopy, Brit. J. Appl. Phys. *7*, 10–12.

Cosslett, V.E. (1969), Energy loss and chromatic aberration in electron microscopy, Z. Angew. Phys. *27*, 138–141.

Cowley, J.M. and A.F. Moodie (1957), The scattering of electrons by atoms and crystals. I. A new theoretical approach, Acta Crystallogr. *10*, 609–619.

Crewe, A.V. (1973), Considerations of specimen damage for the transmission electron microscope, conventional versus scanning, J. Mol. Biol. *80*, 315–325.

Crowther, R.A., D.J. DeRosier and A. Klug (1970a), The reconstruction of a three-dimensional structure from projections and its applications to electron microscopy, Proc. Roy. Soc. Lond. A *317*, 319–340.

Crowther, R.A., L.A. Amos, J.T. Finch, D.J. DeRosier and A. Klug (1970b), Three-dimensional reconstructions of spherical viruses by Fourier synthesis from electron micrographs, Nature *226*, 421–425.

Crowther, R.A., L.A. Amos and J.T. Finch (1975), Three-dimensional image reconstruction of bacteriophages R17 and f2, J. Mol. Biol. *98*, 631–635.

DeRosier, D.J. (1971), The reconstruction of three-dimensional images from electron micrographs, Contemp. Phys. *12*, 437–452.

DeRosier, D.J. and A. Klug (1968), Reconstruction of three-dimensional structures from electron micrographs, Nature *217*, 130–134.

DeRosier, D.J. and P.B. Moore (1970), Reconstruction of three-dimensional images from electron micrographs of structures with helical symmetry, J. Mol. Biol. *52*, 355–369.

Dubochet, J. (1973), High resolution dark-field electron microscopy, J. Microsc. (Oxford) *98*, 334–344.

Frank, J. (1973), Computer processing of electron micrographs, in: Advanced techniques in biological electron microscopy, J.K. Koehler, ed. (Springer-Verlag, Berlin), pp. 215–274.

Gentsch, P., H. Gilde and L. Reimer (1974), Measurement of the top–bottom effect in scanning transmission electron microscopy of thick amorphous specimens, J. Microsc. (Oxford) *100*, 81–92.

Glaeser, R.M. (1975), Radiation damage and biological electron microscopy, in: Physical aspects of electron microscopy, B.M. Siegel and D.R. Beaman, eds. (John Wiley and Sons, New York), pp. 205–229.

Hahn, M., J. Seredynski and W. Baumeister (1976), Inactivation of catalase monolayers by irradiation with 100 keV electrons, Proc. Natl. Acad. Sci. USA *73*, 823–827.

Henderson, R. (1975), The structure of the purple membrane from *Halobacterium halobium*: Analysis of the X-ray diffraction pattern, J. Mol. Biol. *93*, 123–138.

Henderson, R. and P.N.T. Unwin (1975), Three-dimensional model of purple membrane obtained by electron microscopy, Nature *257*, 28–32.

Horne, R.W. (1973), Contrast and resolution from biological objects examined in the electron microscope with particular reference to negatively-stained specimens, J. Microsc. (Oxford) *98*, 286–298.

International Tables for X-ray Crystallography (1969), N.F.M. Henry and K. Lonsdale, eds. (The Kynoch Press, Birmingham), Vol. I. Symmetry groups.

Isaacson, M., D. Johnson and A.V. Crewe (1973), Electron beam excitation and damage of biological molecules; its implications for specimen damage in electron microscopy, Radiat. Res. *55*, 205–224.

Johansen, B.V. (1976), Bright-field electron microscopy of biological specimens. V. A low dose pre-irradiation procedure reducing beam damage, Micron *7*, 145–156.

Klug, A. and R.A. Crowther (1972), Three-dimensional image reconstruction from the viewpoint of information theory, Nature *238*, 435–440.

Klug, A., F.H.C. Crick and H.W. Wyckoff (1958), Diffraction by helical structures, Acta Crystallogr. *11*, 199–213.

Kuo, I.A.M. and R.M. Glaeser (1975), Development of methodology for low exposure, high resolution electron microscopy of biological specimens, Ultramicroscopy *1*, 53–66.

Langer, R., Ch. Poppe, H.J. Schramm and W. Hoppe (1975), Electron microscopy of thin protein crystal sections, J. Mol. Biol. *93*, 159–165.

Lipson, H. and S.G. Lipson (1972), The effect of multiple diffraction on the electron-microscope image, J. Appl. Crystallogr. *5*, 239–240.

Misell, D.L. (1976), On the validity of the weak-phase and other approximations in the analysis of electron microscope images, J. Phys. D: Appl. Phys. *9*, 1849–1866.

Misell, D.L. (1977), Conventional and scanning transmission electron microscopy: image contrast and radiation damage, J. Phys. D: Appl. Phys. *10*, 1085–1107.

Ohlendorf, D.H., M.L. Collins, E.O. Puronen, L.J. Banaszak and S.C. Harrison (1975), Crystalline lipoprotein–phosphoprotein complex in oocytes from *Xenopus laevis*: determination of lattice parameters by X-ray crystallography and electron microscopy, J. Mol. Biol. *99*, 153–165.

Ottensmeyer, F.P., R.F. Whiting and A.P. Korn (1975a), Three-dimensional structure of herring sperm protamine Y–I with the aid of dark field electron microscopy, Proc. Natl. Acad. Sci. USA *72*, 4953–4955.

Ottensmeyer, F.P., R.F. Whiting, E.E. Schmidt and R.S. Clemens (1975b), Electron microtephroscopy of proteins: A close look at the ashes of myokinase and protamine, J. Ultrastruct. Res. *52*, 193–201.

Ottensmeyer, F.P., J.W. Andrew, D.P. Bazett-Jones, A.S.K. Chan and J. Hewitt (1977), Signal-to-noise enhancement in dark field electron micrographs of vasopressin: filtering of arrays of images in reciprocal space, J. Microsc. (Oxford) *109*, 259–268.

Reimer, L. and P. Gentsch (1975), Superposition of chromatic error and beam broadening in transmission electron microscopy of thick carbon and organic specimens, Ultramicroscopy *1*, 1–5.

Serwer, P. (1977), Flattening and shrinkage of bacteriophage T7 after preparation for electron microscopy by negative staining, J. Ultrastruct. Res. *58*, 235–243.

Smith, P.R., T.M. Peters and R.H.T. Bates (1973), Image reconstruction from finite numbers of projections, J. Phys. A: Math., Nucl. Gen. *6*, 361–382.

Smith, P.R., U. Aebi, R. Josephs and M. Kessel (1976), Studies of the structure of the T4 bacteriophage tail sheath-I. The recovery of three-dimensional structural information from the extended sheath, J. Mol. Biol. *106*, 243–275.

Stenn, K. and G.F. Bahr (1970), Specimen damage caused by the beam of the transmission electron microscope, a correlative reconsideration, J. Ultrastruct. Res. *31*, 526–550.

Ten Eyck, L.F. (1973), Crystallographic fast Fourier transforms, Acta Crystallogr. *A29*, 183–191.

Unwin, P.N.T. (1974), Electron microscopy of the stacked disk aggregate of tobacco mosaic virus protein. II. The influence of electron irradiation on the stain distribution, J. Mol. Biol. *87*, 657–670.

Unwin, P.N.T. (1975), Beef liver catalase structure: interpretation of electron micrographs, J. Mol. Biol. *98*, 235–242.

Unwin, P.N.T. and R. Henderson, (1975), Molecular structure determination by electron microscopy of unstained crystalline specimens, J. Mol. Biol. *94*, 425–440.

Vaĭnshteĭn, B.K. (1973), Three-dimensional electron microscopy of biological macromolecules, Sov. Phys.-Usp. *16*, 185–206 (English translation).

Van Holde, K.E. (1971), Physical biochemistry (Prentice-Hall, Inc., Engelwood Cliffs, New Jersey).

Williams, R.C. and H.W. Fisher (1970), Electron microscopy of tobacco mosaic virus under conditions of minimal beam exposure, J. Mol. Biol. *52*, 121–123.

O passi graviora, dabit deus his quoque finem.   (Virgil, Aeneid)
(You have endured worse things, God will grant an end even to these.)

Appendix

List of suppliers

1. Auxiliary viewing screens – phosphors (minimum radiation techniques)

Levy West Laboratories
Millmarsh Lane
Brimsdown
Middlesex EN5 7QW
U.K.

JEOL Ltd
1418 Nakagami Akishima
Tokyo 196
Japan

JEOL (U.K.) Ltd.
JEOL House
Grove Park
Edgeware Road
Colindale
London NW9 0JN
U.K.

JEOL (U.S.A.) Inc.
477 Riverside Avenue
Medford
Massachusetts 02155
U.S.A.

2. Catalase test grids (optical diffractometer calibration)

Agar Aids
66a Cambridge Road
Stansted
Essex CM24 8DA
U.K.

3. Catalase suspension (optical diffractometer calibration)

The Boehringer Corporation (London) Ltd.
Bell Lane
Lewes
East Sussex BN7 1LG
U.K.

Boehringer Mannheim Corporation
219 East 44th Street
New York
New York 10017
U.S.A.

4. Copper gratings (optical diffractometer calibration)

Ernest Fullam Inc.
P.O. Box 444
Schenectady
New York 12301
U.S.A.

Graticules Ltd.
Sovereign Way
Tonbridge
Kent TN19 1RN
U.K.

5. Electron microscope accessories (including thin film apertures)

Agar Aids (*see* 2)

Electron Microscope Aids
6 Lime Trees
Christian Malford
Chippenham
Wiltshire SN15 4BN
U.K.

Extech International Corporation
177 State Street
Boston
Massachusetts 02109
U.S.A.

Ernest Fullam Inc. (*see* 4)

Graticules Ltd. (*see* 4)

Ladd Research Industries Inc.
P.O. Box 901
Burlington
Vermont 05401
U.S.A.

Polaron Equipment Ltd.
60/62 Greenhill Crescent
Holywell Industrial Estate
Watford
Hertfordshire WD1 8RL
U.K.

International Enzymes Ltd.
(U.K. agents for Polysciences)
Hanover Way
Vale Road
Windsor
Berkshire SL4 5NJ
U.K.

Polysciences Inc.
Paul Valley Industrial Estate
Warrington
Pennsylvania 18976
U.S.A.

Taab Laboratories
52 Kidmore End Road
Emmer Green
Reading
Berkshire RG4 8SE
U.K.

6. Electron emulsions (density – exposure characteristics)

Ilford Ltd.
23 Roden Street
Ilford
Essex IG1 2AB
U.K.

Kodak Ltd.
P.O. Box 33
Hemel Hempstead
Hertfordshire HP2 7EU
U.K.

Eastman Kodak Co.
Rochester
New York 14650
U.S.A.

Information on EM emulsions:

Eastman Kodak Co.
Electron Microscopy and Photography
    P-236
Kodak Electron Image Plates P-116
Kodak Materials for Electron Micrography P-204

7. Lasers (for optical diffractometers)

Scientifica and Cook Electronics Ltd.
78 Bollo Bridge Road
Acton
London W3 8AU
U.K.

The Ealing Corporation
(U.S.A. agents for Scientifica and Cook
Electronics Ltd.)
South Natick
Massachusetts
U.S.A.

Spectra – Physics
1250 West Middlefield Road
Mountain View
California 94040
U.S.A.

Spectra – Physics Ltd.
17 Brick Knoll Park
St. Albans
Hertfordshire AL1 5BR
U.K.

8. Optical components
(lenses, mounts, benches)

Ealing Beck Ltd.
Greycaine Road
Watford
Hertfordshire WD2 4PW
U.K.

The Ealing Corporation (*see* 7)

George Elliot and Sons Ltd.
Worcester House
Vintners Place
London EC4V 3MM
U.K.

Oriel Optics Corporation
1 Market Street
Stamford
Connecticut 06902
U.S.A.

The Precision Tool and Instrument Co. Ltd.
353 Bensham Lane
Thornton Heath
Surrey CR4 7ER
U.K.

Spectra – Physics Ltd. (*see* 7)

9. Optical diffractometers – image analysis systems

Polaron Equipment Ltd. (*see* 5)

Rank Precision Instruments
P.O. Box 36
Guthlaxton Street
Leicester LE2 OSP
U.K.

10. Scanning densitometers (for digitising micrographs)

Agar Aids (*see* 2)
(low resolution rotating drum type)

Cassidy Scott Instruments
(U.K. agents for Optronics International
Sales Corporation)
12 Weston Avenue
Whickham
Newcastle upon Tyne NE16 5TS
U.K.

Joyce-Loebl Ltd.
(U.K. agents for Perkin-Elmer Corporation)
Team Valley
Princessway
Gateshead NE11 OUJ
U.K.

Optronics International Sales Corporation
7 Stuart Road
Chelmsford
Massachusetts 01824
U.S.A.
(mainly rotating drum type)

Perkin-Elmer Corporation
Boller and Chivens Division
916 Meridian Avenue
South Pasadena
California 91030
U.S.A.
(high resolution flat-bed type)

11. Computer processing systems (television and display facilities)

Comtal Corporation
333 North Santa Anita Avenue
Arcadia
California 91006
U.S.A.

Spatial Data Systems Inc.
(Publisher of free booklet 'Computer Eye',
a handbook of image processing, 1975)
P.O. Box 249
508 S. Fairview Avenue
Goleta
California 93017
U.S.A.

Subject index